Ham Radio Study Guide:
Manual for Technician Class, General Class, and Amateur Extra Class

Copyright © 2015 by Trivium Test Prep

ALL RIGHTS RESERVED. By purchase of this book, you have been licensed one copy for personal use only. No part of this work may be reproduced, redistributed, or used in any form or by any means without prior written permission of the publisher and copyright owner. Questions regarding permissions or copyrighting may be sent to support@triviumtestprep.com.

TABLE OF CONTENTS

1. About Trivium Test Prep and How to Use this Guide - - 5
2. Introduction - - - - - - - 7
3. Licensing Overview - - - - - - 9
4. Technician Class (Element 2) - - - - - 11
    - 4.1. Subelement T1 - - - - - 13
    - 4.2. Subelement T2 - - - - - 19
    - 4.3. Subelement T3 - - - - - 23
    - 4.4. Subelement T4 - - - - - 26
    - 4.5. Subelement T5 - - - - - 29
    - 4.6. Subelement T6 - - - - - 33
    - 4.7. Subelement T7 - - - - - 37
    - 4.8. Subelement T8 - - - - - 41
    - 4.9. Subelement T9 - - - - - 45
    - 4.10. Subelement T0 - - - - - 47
    - 4.11. Study Tips - - - - - - 51
5. General Class (Element 3) - - - - - 53
    - 5.1. Subelement G1 - - - - - 54
    - 5.2. Subelement G2 - - - - - 59
    - 5.3. Subelement G3 - - - - - 65
    - 5.4. Subelement G4 - - - - - 68
    - 5.5. Subelement G5 - - - - - 73
    - 5.6. Subelement G6 - - - - - 77
    - 5.7. Subelement G7 - - - - - 81
    - 5.8. Subelement G8 - - - - - 84
    - 5.9. Subelement G9 - - - - - 86

- 5.10. Subelement G0 — 91
- 5.11. Study Tips — 94
- 6. Amateur Extra Class (Element 4) — 95
  - 6.1. Subelement E1 — 96
  - 6.2. Subelement E2 — 104
  - 6.3. Subelement E3 — 109
  - 6.4. Subelement E4 — 112
  - 6.5. Subelement E5 — 118
  - 6.6. Subelement E6 — 125
  - 6.7. Subelement E7 — 132
  - 6.8. Subelement E8 — 144
  - 6.9. Subelement E9 — 149
  - 6.10. Subelement E0 — 160
  - 6.11. Study Tips — 162
- 7. APPENDIX 1: Element 2 Practice Exam — 163
  - 7.1. Answer Key — 169
- 8. APPENDIX 2: Element 3 Practice Exam — 171
  - 8.1. Answer Key — 177
- 9. APPENDIX 3: Element 4 Practice Exam — 179
  - 9.1. Answer Key — 187
- 10. APPENDIX 4: Resource Directory — 189

ABOUT TRIVIUM TEST PREP

Trivium Test Prep uses industry professionals with decades worth of knowledge in the fields they have mastered, proven with degrees and honors in law, medicine, business, education, military, and more to produce high-quality test prep books such as this for students.

Our study guides are specifically designed to increase ANY student's score, regardless of his or her current scoring ability. At only 25% - 35% of the page count of most study guides, you will increase your score, while significantly decreasing your study time.

HOW TO USE THIS GUIDE

This guide is not meant to reteach you material you have already learned or waste your time on superfluous information. We hope you use this guide to focus on the key concepts you need to master for the test and develop critical test-taking skills. To support this effort, the guide provides:

- Practice questions with worked-through solutions
- Key test-taking tactics that reveal the tricks and secrets of the test
- Simulated one-on-one tutor experience
- Organized concepts with detailed explanations
- Tips, tricks, and test secrets revealed

Because we have eliminated "filler" or "fluff", you will be able to work through the guide at a significantly faster pace than other prep books. By allowing you to focus ONLY on those concepts that will increase your score, study time is more effective and you are less likely to lose focus or get mentally fatigued.

1. Introduction

Whether you are delving into ham radio for the first time or looking to move up in rank among licensed operators, this study guide is a great starting point. Many would-be operators aspire to advance in amateur radio but struggle in preparing for the licensing exam because they simply don't know what to study. This manual will help you avoid that problem.

Our goal is to provide aspiring operators with a concise study aid that will help those prospective licensees pass the required test. It is neither necessary nor practical to supply you with an all-inclusive reference of rules, regulations, operational practices, and/or technical theory. Such conventional resources, both free and for purchase, are readily available. And we'll show you when, where, and how to leverage those resources to your advantage.

But whatever your licensing objective, this study guide is key to passing the related exam. Although nothing else is needed, we'll also show you additional techniques and tools that can skyrocket your knowledge to an expert level. Here's what you'll find inside the manual:

- An overview of the amateur (ham) radio license classifications and test requirements

- Detailed rundowns for each exam syllabus
- Comprehensive listings of test notes to answer every potential exam question
- Technical notes for every diagram used on all licensing exams
- A full-length practice test for each amateur radio license examination
- Additional resources for getting more practice and expanding your knowledge base

Read the entire manual from beginning to end if you like, or simply review the sections that suit your needs. The choice is yours. Either way, you are sure to benefit from the accurate and updated information contained within. Conceived, designed, and assembled by industry experts, this compendium's sole purpose is to help you pass any or all of the ham radio licensing exams. And you won't find another learning tool as complete and concise as this guide, anywhere – we know because we looked and could not find one.

So if you are ready to take that next step in the exciting world of amateur (ham) radio and pass your license exam with confidence, then grab your favorite drink, sit back, relax, and let's get started...

2. LICENSING OVERVIEW

"The amateur and amateur-satellite services are for qualified persons of any age who are interested in radio technique solely with a personal aim and without pecuniary interest. These services present an opportunity for self-training, intercommunication, and technical investigations."
— *Federal Communications Commission*

The Federal Communications Commission (FCC) currently issues amateur radio licenses for three operator classifications: *Technician*, *General*, and *Amateur Extra*. All new operators start with the Technician Class, then advance to the General and Amateur Extra classes by successfully passing the applicable written examination elements: 2, 3, and 4 respectively.

Volunteer Examiners (VEs) will prepare and administer each test based on predetermined question pools. Credits for successfully passing previous tests are awarded to operators, eliminating the need to repeat exams when advancing to the next operator classification. Here is a summary of all three current operator classes and their respective test requirements:

Technician

Licensed Technician Class operators are given privileges on 4 amateur service bands in the HF range (3 – 30 MHz), and may also operate on channels in 17 frequency bands above 50 MHz. Passing a Technician Class written exam (Element 2) requires a minimum score of 26 correct answers from a total of 35 questions.

General

General Class licensees are authorized privileges on all 27 Amateur Radio Service bands. General Class written exams (Element 3) are also 35 questions in length and again require a minimum passing score of 26 correct answers.

Amateur Extra

The Amateur Extra Class operator privileges include additional HF service band spectrum. Amateur Extra Class written exams (Element 4) include a total of 50 questions, from which a minimum passing score of 37 correct answers is required.

3. Technician Class (Element 2)

Technician Class written examinations are referred to as "Element 2." Topics covered during the examinations include: FCC rules, operator responsibilities, operating procedures, radio wave characteristics, station set up and practices, electrical principles and components, station equipment, modes of modulation, antennas and feed lines, and electrical safety.

Element 2 exams are derived from 10 subelements containing 35 targeted question groups. The current public question pool contains a total of 423 available test questions. Element 2 tests are made up of 35 random questions, of which at least 26 must be answered correctly to qualify for a Technician Class operator license. Following is a rundown of applicable subelements and related question groups.

3.1 SUBELEMENT T1

Subelement T1 contains six targeted question groups with a total of 76 potential test questions. Six of the 35 [Element 2] test questions will be drawn from subelement T1.

Group T1A – Amateur Radio Service

Questions in group T1A cover the following topics: location and purpose of FCC rules, basic terms, purpose of amateur services, permissible use, interference, and spectrum management.

T1A Test Notes

- One purpose of the Amateur Radio Service is for advancing both technical and communication skills.
- In the United States, rules for the Amateur Radio Service are regulated and enforced by the FCC.

- Rules governing the Amateur Radio Service are located in the FCC Code of Federal Regulations (CFR), Title 47, and Part 97.
- The FCC defines "harmful interference" as "that which seriously degrades, obstructs, or repeatedly interrupts a radio communication service operating in accordance with the Radio Regulations".
- Another purpose of the Amateur Radio Service is for enhancing international goodwill.
- Radionavigation Service is ALWAYS protected from interference by amateur signals.
- The FCC defines "telemetry" as "a one-way transmission of measurements at a distance from the measuring instrument".
- Frequency coordinators recommend the transmit/receive channels plus other parameters for auxiliary and repeater stations.
- Frequency coordinators are chosen by amateur operators of a local or regional area whose stations are eligible to be auxiliary or repeater stations.
- The FCC defines an "amateur station" as "a station in the Amateur Radio Service consisting of the apparatus necessary for carrying on radio communications".
- Willful interference to another amateur radio station is NEVER permitted.
- Allowing someone to conduct radio experiments and communicate with other licensed hams around the world is considered a permissible use of the Amateur Radio Service.
- The FCC defines "telecommand" as "a one-way transmission to initiate, modify, or terminate functions of a device at a distance".
- If interfering with a radiolocation station outside the United States, you MUST stop operating or take steps to eliminate the harmful interference.

Group T1B – Authorized Frequencies

Questions in group T1B cover the following topics: ITU regions, frequency allocations, emission modes, spectrum sharing, restricted sub-bands, and operating near band edges.

T1B Test Notes

- The International Telecommunication Union (ITU) is an agency of the United Nations responsible for information and communication technology issues.
- Some U.S. territories are located in ITU regions other than region 2, and frequency assignments may therefore differ from those of the 50 U.S. states.
- 52.525 MHz falls within the 6 meter band.
- 146.52 MHz falls within the 2 meter band.
- Licensed Technician Class operators in ITU region 2 are authorized to use 443.350 MHz in the 70 cm band.
- Licensed Technician Class operators are authorized to use 1296 MHz in the 23 cm band.
- 223.50 MHz falls within the 1.25 meter band.
- Because amateur service is secondary in some portions of the 70 cm band, U.S. amateurs may find non-amateur stations in the bands and must avoid interfering with them.
- Avoid setting transmit frequencies at the edge of amateur bands and sub-bands, so modulation sidebands do not extend beyond the band edge and to allow for transmitter frequency drift or calibration error.
- 6 meter, 2 meter, and 1.25 meter bands all have mode-restricted sub-bands.
- Only continuous wave emission (CW) is permitted in the mode-restricted sub-bands at 50.0 to 50.1 MHz and 144.0 to 144.1 MHz.
- Amateur frequency assignments for U.S. stations operating maritime mobile are not the same everywhere in the world because assignments vary among the three ITU regions.
- Data emission may be used between 219 and 220 MHz.

Group T1C – Operator Licensing

Questions in group T1C cover the following topics: operator classes, special event and vanity call sign systems, international communications, reciprocal operation, station licensing, where the FCC regulates amateur service, FCC license database info, license term, renewal, and grace period.

T1C Test Notes

- A special event call sign has a single letter in both its prefix and suffix.
- W3ABC is an example of a valid U.S. amateur radio station call sign.
- Communications incidental to the purposes of the amateur service and remarks of a personal character are permitted by an FCC-licensed amateur station.
- You are allowed to operate your amateur station in a foreign country only when authorized by the foreign country.
- K1XXX is an example of a vanity call sign available to Technician Class operators.
- In addition to places where communications are regulated by the FCC, licensed amateur stations may transmit from any vessel or craft located in international waters and documented or registered in the United States.
- Failure to provide the FCC with the correct mailing address could result in revocation of the station license or suspension of the operator license.
- The normal term for an FCC-issued primary station/operator amateur radio license grant is ten years.
- The grace period following the expiration of an amateur license within which the license may be renewed is two years.
- After passing the examination for your first amateur radio license, you may operate a transmitter on an amateur service frequency as soon as your operator/station license grant appears in the FCC license database.
- If your license expires, even when still within the allowable grace period, transmitting is not allowed until the FCC license database shows that the license has been renewed.
- Any licensed amateur may select a desired call sign under the vanity call sign rules.
- Technician, General, and Amateur Extra are the new license classifications currently available from the FCC.

- Only the person named as trustee on the club station license grant may select a vanity call sign for a club station.

Group T1D – *Authorized and Prohibited Transmission*

Questions in group T1D cover the following topics: international communications, exchange of information between services, music, indecent language, compensation issues, retransmission, codes and ciphers, equipment sales, unidentified transmissions, and broadcasting.

T1D Test Notes

- Amateur stations are prohibited from exchanging communications with any country whose administration has notified the ITU that it objects to such communications.

- Amateur radio stations may exchange messages with a U.S. military station during an Armed Forces Day Communications Test.

- Transmission of codes or ciphers that hide the meaning of a message is allowed only when transmitting control commands to space stations or radio control craft.

- The only time an amateur station is authorized to transmit music is when incidental to an authorized retransmission of manned spacecraft communications.

- Amateur radio operators may use their stations to notify other amateurs of the availability of equipment for sale or trade when the equipment is normally used in an amateur station, and such activity is not conducted on a regular basis.

- Language that may be considered indecent or obscene is strictly prohibited.

- Amateur auxiliary, repeater, or space stations can automatically retransmit the signals of other amateur stations.

- The control operator of an amateur station may receive compensation for operating the station when the communication is incidental to classroom instruction at an educational institution.

- Assuming no other means is available, amateur stations are authorized to transmit signals related to broadcasting, program production or news gathering only where such communications directly relate to the immediate safety of human life or protection of property.
- According to the FCC, the term "broadcasting" refers "to transmissions intended for reception by the general public".
- An amateur station may transmit without identifying when transmitting signals to control a model craft.
- Amateur radio stations may engage in broadcasting when transmitting code practice, information bulletins, or transmissions necessary to provide emergency communications.

Group T1E – Control Operator and Control Types

Questions in group T1E cover the following topics: requirements, eligibility, designation, privileges and duties, local control points, automatic and remote control, and location.

T1E Test Notes

- Amateur stations are NEVER permitted to transmit without a control operator.
- Only a person for whom an amateur operator/primary station license grant appears in the FCC database or who is authorized for alien reciprocal operation may be designated as the control operator of an amateur station.
- The station licensee must designate the station control operator.
- The transmitting privileges of an amateur station are determined by the class of operator license held by the control operator.
- A station control point is the location at which the control operator function is performed.
- Automatic packet reporting system (APRS) network digipeaters operate under automatic control.
- When the control operator is not the station licensee, the control operator and the station licensee are equally responsible for proper operation of the station.

- Repeater operation is an example of automatic control.
- Local control is when the control operator is at the control point.
- Operating a station over the Internet is an example of remote control.
- Unless documentation to the contrary is in the station records, the FCC presumes the station licensee to be the control operator of an amateur station.
- Normally, at no time may a Technician Class licensee be the control operator of a station operating in an exclusive Extra Class operator segment of the amateur bands.

Group T1F – Miscellaneous

Questions in group T1F cover the following topics: station identification, repeaters, third-party communications, club stations, and FCC inspection.

T1F Test Notes

- A tactical call sign is the type of identification used when identifying a station on the air as Race Headquarters.
- When using tactical identifiers such as "Race Headquarters" during a community service net operation, the station must transmit its FCC-assigned call sign at the end of each communication and every 10 minutes during a communication.
- Amateur stations are required to transmit their assigned call signs at least every 10 minutes during and at the end of a communication.
- English is recognized as an acceptable language to use for station identification when operating in a phone sub-band.
- Using CW or phone emission is a required method of call sign identification for a station transmitting phone signals.
- KL7CC stroke W3, KL7CC slant W3, and KL7CC slash W3 are all acceptable formats of a self-assigned indicator when identifying using a phone transmission.
- When a non-licensed person is allowed to speak to a foreign station using a station under the control of a Technician Class control operator, the

foreign station must be one with which the United States has a third-party agreement.

- /KT, /AE, or /AG is required to be transmitted after a station call sign when using new license privileges earned by Certificate of Successful Completion of Examination (CSCE) while waiting for a license upgrade to appear in the FCC license database.
- A repeater station simultaneously retransmits the signal of another amateur station on a different channel or channels.
- The control operator of the originating station is accountable if a repeater inadvertently retransmits communications that violate FCC rules.
- The FCC authorizes the transmission of non-emergency third-party communications to any foreign station whose government permits such communications.
- At least four persons are required to be members of a club for a club station license to be issued by the FCC.
- A station licensee must make the station and its records available for FCC inspection at any time upon request by an FCC representative.

3.2 SUBELEMENT T2

Subelement T2 contains three targeted question groups with a total of 37 potential test questions. Three of the 35 [Element 2] test questions will be drawn from subelement T2.

Group T2A – Station Operation

Questions in group T2A cover the following topics: choosing operating frequencies, calling other stations, test transmissions, procedural signs, using minimum power, band plans, calling frequencies, and repeater offsets.

T2A Test Notes

- Plus or minus 600 kHz is the most common repeater frequency offset in the 2 meter band.
- 446.000 MHz is the national calling frequency for frequency modulation (FM) simplex operations in the 70 cm band.
- Plus or minus 5 MHz is a common repeater frequency offset in the 70 cm band.
- If you know another station's call sign, an appropriate way to call it is to say their call sign and then identify with your call sign.
- Respond to a station calling CQ by transmitting the other station's call sign followed by your call sign.
- An operator must properly identify the transmitting station when making on-air transmissions to test equipment or antennas.
- Station identification is required at least every 10 minutes during the test and at the end of the test.
- The procedural signal CQ means "Calling any station."
- Your call sign is often transmitted in place of "CQ" to indicate that you are listening on a repeater.
- Beyond the privileges established by the FCC, a band plan is a voluntary guideline for using different modes or activities within an amateur band.

- Under normal (non-distress) circumstances, use the minimum power necessary to carry out the desired communication, and do not exceed the maximum power permitted on a given band.
- When choosing an operating frequency for calling CQ, make sure you are in your assigned band, listen first to be sure that no one else is using the frequency, and ask if the frequency is in use.

Group T2B – VHF/UHF Operating Practices

Questions in group T2B cover the following topics: SSB phones, FM repeaters, simplex, splits and shifts, CTCSS, DTMF, tone squelch, carrier squelch, phonetics, problem resolution, and Q signals.

T2B Test Notes

- Simplex communication is the term used to describe an amateur station that is transmitting and receiving on the same frequency.
- Continuous tone coded squelch system (CTCSS) is the term used to describe use of a sub-audible tone transmitted with normal voice audio to open the squelch of a receiver.
- Carrier squelch describes the muting of receiver audio controlled solely by the presence or absence of an RF signal.
- You might be able to hear but not access a repeater even when transmitting with the proper offset if the receiver requires an audio tone burst, a CTCSS tone, or a digital code squelch (DCS) tone for access.
- The amplitude of a modulating signal determines the amount of deviation of an FM signal.
- When the deviation of an FM transmitter is increased, its signal occupies more bandwidth.
- If your microphone gain is too high, it could result in overdeviation and your FM signal could interfere with stations on nearby frequencies.
- When two stations are transmitting on the same frequency common courtesy should prevail, but no one has absolute right to an amateur frequency.

- Use of a phonetic alphabet is encouraged by the FCC while identifying your station when using a phone.
- A QRM signal indicates that you are receiving interference from other stations.
- A QSY signal indicates that you are changing frequency.
- Consider communicating via simplex rather than a repeater when the stations can communicate directly without using a repeater.
- Use of SSB phone is permitted in at least some portion of all amateur service bands above 50 MHz.

Group T2C – Public Service

Questions in group T2C cover the following topics: emergency and non-emergency operations, applicability of FCC rules, RACES and ARES, net and traffic procedures, plus emergency restrictions.

T2C Test Notes

- FCC rules ALWAYS apply to the operation of an amateur station.
- If commercial power is out, one way to recharge a 12-volt lead-acid station battery is to connect the battery in parallel with a vehicle's battery and run the engine.
- To ensure that the receiving station copies voice message traffic containing proper names and unusual words correctly, such words and terms should be spelled out using a standard phonetic alphabet.
- Both the Radio Amateur Civil Emergency Service (RACES) and Amateur Radio Emergency Service (ARES) may provide communications during emergencies.
- RACES is an emergency service using amateur operators, stations, and frequencies for emergency management or civil defense communications; operators are certified by a civil defense organization.
- Begin your transmission by saying "priority" or "emergency" followed by your call sign to get the immediate attention of a net control station (NCS) when reporting an emergency.

- If checked into an emergency traffic net, remain on frequency without transmitting until asked to do so by the net control station.
- Passing messages exactly as received is a characteristic of good emergency traffic handling.
- Amateur station control operators are permitted to operate outside the frequency privileges of their license class ONLY if necessary in situations involving the immediate safety of human life or protection of property.
- The preamble in a formal traffic message is the information needed to track the message as it passes through the amateur radio traffic handling system.
- In reference to a formal traffic message, the "check" is a count of the number of words or word equivalents in the text portion of the message.
- ARES is comprised of licensed amateurs who have voluntarily registered their qualifications and equipment for communications duty in the public service.

3.3 SUBELEMENT T3

Subelement T3 contains three targeted question groups with a total of 34 potential test questions. Three of the 35 [Element 2] test questions will be drawn from subelement T3.

Group T3A – Radio Wave Characteristics

Questions in group T3A cover the following topics: how radio waves travel, fading, multipath, wavelength vs. penetration, plus antenna orientation.

T3A Test Notes

- If another operator reports that your station's 2 meter signals were strong just a moment ago but are now weak or distorted, try moving a few feet or changing the direction of your antenna because reflections may be causing multipath distortion.
- UHF signals are often more effective from inside buildings than VHF signals because the shorter wavelength allows them to more easily penetrate the structure of buildings.

- Horizontal antenna polarization is normally used for long-distance, weak-signal CW and SSB contacts using the VHF and UHF bands.
- Signals could be significantly weaker if the antennas at opposite ends of a VHF or UHF line of sight radio link are not using the same polarization.
- If obstructions are blocking the direct line of sight path when using a directional antenna, try accessing a distant repeater by finding a path that reflects signals to the repeater.
- "Picket fencing" describes the rapid fluttering sound sometimes heard from mobile stations that are moving while transmitting.
- Electromagnetic waves carry radio signals between transmitting and receiving stations.
- Irregular fading of signals received by ionospheric reflection are likely caused by the random combining of signals arriving via different paths.
- Skip signals refracted from the ionosphere are elliptically polarized; therefore, either vertically or horizontally polarized antennas may be used for transmission or reception.
- Error rates are likely to increase if data signals propagate over multiple paths.
- The ionosphere enables propagation of radio signals around the world.

Group T3B – Radio and Electromagnetic Wave Properties

Questions in group T3B cover the following topics: electromagnetic spectrum, wavelength vs. frequency, velocity of electromagnetic waves, and calculating wavelength.

T3B Test Notes

- Wavelength is the distance a radio wave travels during one complete cycle.
- The orientation of the electric field is used to describe radio wave polarization.
- Electric and magnetic fields are the two components of a radio wave.
- Radio waves travel through free space at the speed of light.
- The wavelength of a radio wave gets shorter as the frequency increases.

- Wavelength in meters equals 300 divided by frequency in megahertz.
- The approximate wavelength of radio waves is often used to identify the different frequency bands.
- 30 to 300 MHz are the frequency limits of the VHF spectrum.
- 300 to 3000 MHz are the frequency limits of the UHF spectrum.
- The 3 to 30 MHz frequency range is referred to as HF.
- 300,000,000 meters per second is the approximate velocity of a radio wave as it travels through free space.

Group T3C – Propagation Modes

Questions in group T3C cover the following topics: line of sight, sporadic E, meteor and auroral scatter and reflections, tropospheric ducting, F layer skip, plus radio horizon.

T3C Test Notes

- Direct UHF signals (not via a repeater) rarely get heard from stations outside a local coverage area because UHF signals are usually not reflected by the ionosphere.
- When VHF signals are received from long distances, the signals are being refracted from a sporadic E layer.
- VHF signals received via auroral reflection exhibit rapid fluctuations of strength and often sound distorted.
- Sporadic E propagation is most commonly associated with occasional strong over-the-horizon signals on the 10, 6, and 2 meter bands.
- Knife-edge diffraction effect might cause radio signals to be heard despite obstructions between the transmitting and receiving stations.
- Tropospheric scatter is responsible for allowing over-the-horizon VHF and UHF communications to ranges of approximately 300 miles on a regular basis.
- The 6 meter band is best suited for communicating via meteor scatter.
- Temperature inversions in the atmosphere cause tropospheric ducting.

- From dawn to shortly after sunset during periods of high sunspot activity is generally the best time for long-distance 10 meter band propagation via the F layer.
- Radio horizon is the distance over which two stations can communicate by direct path.
- VHF and UHF radio signals usually travel somewhat farther than the visual line of sight distance between two stations because the Earth seems less curved to radio waves than to light.
- The 6 or 10 meter bands may provide long-distance communications during the peak of a sunspot cycle.

3.4 SUBELEMENT T4

Subelement T4 contains two targeted question groups with a total of 24 potential test questions. Two of the 35 [Element 2] test questions will be drawn from subelement T4.

Group T4A – Station Setup

Questions in group T4A cover the following topics: microphones, reducing unwanted emissions, power sources, connecting computers, RF grounding, connecting digital equipment, and connecting SWR meters.

T4A Test Notes
- Some amateur transceiver microphone connectors include push-to-talk and voltages for powering the microphone.
- A computer can be used as part of an amateur radio station for logging contacts and contact information, generating and decoding digital signals, plus sending and/or receiving CW.
- A good reason to use regulated power supplies for communications equipment is that they prevent voltage fluctuations from reaching sensitive circuits.

- A filter must be installed between the transmitter and antenna to reduce harmonic emissions from a station.
- An in-line SWR meter must be connected in series with the feed line between the transmitter and antenna to monitor the standing-wave ratio of a station antenna system.
- A terminal node controller is connected between the transceiver and computer in a packet radio station.
- When conducting digital communications, a computer sound card provides audio to the microphone input and converts received audio to digital form.
- Flat strap is the best conductor to use for RF grounding.
- You can use a ferrite choke to cure distorted audio caused by RF current flowing on the shield of a microphone cable.
- A vehicle's alternator is the source of high-pitched whining that varies with engine speed in a mobile transceiver's receive audio.
- The negative return connection of a mobile transceiver's power cable should be connected at the battery or engine block ground strap.
- If another operator reports a variable high-pitched whine on the audio from your mobile transmitter, it is possible that noise on the vehicle's electrical system is being transmitted along with your speech audio.

Group T4B – Operating Controls

Questions in group T4B cover the following topics: tuning, filters, squelch, automatic gain control (AGC), repeater offset, and memory channels.

T4B Test Notes

- The output signal might become distorted if a transmitter is operated with the microphone gain set too high.
- The keypad or variable-frequency oscillator (VFO) knob can be used to enter the operating frequency on a modern transceiver.
- The purpose of the squelch control on a transceiver is to mute receiver output noise when no signal is being received.

- Storing the frequency in a memory channel will enable quick access to a favorite frequency on your transceiver.
- Turning on the noise blanker can reduce ignition interference to a receiver.
- The receiver RIT or clarifier can be used if the voice pitch of a single-sideband signal seems too high or low.
- The term "RIT" stands for Receiver Incremental Tuning.
- Having multiple receive bandwidth choices on a multimode transceiver permits noise or interference reduction by selecting a bandwidth matching the mode.
- 2400 Hz is an appropriate receive filter bandwidth to select in order to minimize noise and interference for SSB reception.
- 500 Hz is an appropriate receive filter bandwidth to select in order to minimize noise and interference for CW reception.
- Repeater offset is the difference between a repeater's transmit and receive frequencies.
- Automatic gain control (AGC) is for keeping received audio relatively constant.

3.5 SUBELEMENT T5

Subelement T5 contains four targeted question groups with a total of 50 potential test questions. Four of the 35 [Element 2] test questions will be drawn from subelement T5.

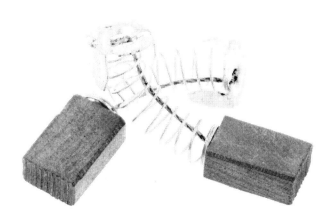

Group T5A – Electrical Principles

Questions in group T5A cover the following topics: units and measures, basic terms, voltage and current, conductors and insulators, plus alternating and direct current.

T5A Test Notes

- Electrical current is measured in amperes.
- Electrical power is measured in watts.
- The flow of electrons in an electric circuit is known as current.
- Direct current flows only in one direction.
- Voltage is the electromotive force (EMF) that causes electron flow.
- Mobile transceivers usually require about 12 volts.
- Copper is a good electrical conductor.
- Glass is a good electrical insulator.

- Alternating current reverses direction on a regular basis.
- Power is the rate at which electrical energy is used.
- The volt is a basic unit of electromotive force.
- Frequency is the number of times per second that an alternating current reverses direction.

Group T5B – Math for Electronics

Questions in group T5B cover the conversion of electrical units, decibels, and the metric system.

T5B Test Notes

- 1.5 amperes are equal to 1500 milliamperes.
- A radio signal frequency of 1,500,000 hertz can also be expressed as 1500 kHz.
- One kilovolt is equal to one thousand volts.
- One microvolt is equal to one one-millionth of a volt.
- 0.5 watts is equal to 500 milliwatts.
- 3 amperes are equal to 3000 milliamperes.
- 3.525 MHz is equal to 3525 kHz.
- 1 microfarad is equal to 1,000,000 picofarads.
- The approximate amount of change, measured in decibels (dB), of a power increase from 5 watts to 10 watts equals 3 dB.
- The approximate amount of change, measured in decibels (dB), of a power decrease from 12 watts to 3 watts equals -6 dB.
- The approximate amount of change, measured in decibels (dB), of a power increase from 20 watts to 200 watts equals 10 dB.
- 28.400 MHz is equal to 28,400 kHz.
- 2.425 GHz is equal to 2425 MHz.

Group T5C – Electronic Principles

Questions in group T5C cover the following topics: capacitance, inductance, current flow in circuits, alternating current, the definition of RF, DC power calculations, and impedance.

T5C Test Notes

- Capacitance is the ability to store energy in an electric field.
- The farad is a basic unit of capacitance.
- Inductance is the ability to store energy in a magnetic field.
- The henry is a basic unit of inductance.
- The hertz is a unit of frequency.
- The abbreviation "RF" refers to radio frequency signals of all types.
- Radio waves are electromagnetic waves that travel through space.
- The formula used to calculate electrical power in a DC circuit is: power (P) equals voltage (E) multiplied by current (I)
- A circuit with an applied voltage of 13.8 volts DC and 10 amperes of current uses 138 watts of power.
- A circuit with an applied voltage of 12 volts DC and 2.5 amperes of current uses 30 watts of power.
- A circuit with an applied voltage of 12 volts DC and a 120-watt load will draw 10 amperes of current.
- Impedance is a measure of the opposition to AC current flow in a circuit.
- The ohm is a unit of impedance.

Group T5D – Ohm's Law

Questions in group T5D focus only on the formulas and usage for Ohm's Law.

T5D Test Notes

- The formula used to calculate current in a circuit is: current (I) equals voltage (E) divided by resistance (R)
- The formula used to calculate voltage in a circuit is: voltage (E) equals current (I) multiplied by resistance (R)
- The formula used to calculate resistance in a circuit is: resistance (R) equals voltage (E) divided by current (I)
- The resistance of a circuit with an applied voltage of 90 volts and 3 amperes of current equals 30 ohms.

- The resistance in a circuit with an applied voltage of 12 volts and 1.5 amperes of current equals 8 ohms.
- The resistance of a circuit with an applied voltage of 12 volts and 4 amperes of current equals 3 ohms.
- The current flow in a circuit with an applied voltage of 120 volts and a resistance of 80 ohms equals 1.5 amperes.
- The current flowing through a 100-ohm resistor that is connected across 200 volts equals 2 amperes.
- The current flowing through a 24-ohm resistor that is connected across 240 volts equals 10 amperes.
- The voltage across a 2-ohm resistor with 0.5 amperes of current equals 1 volt.
- The voltage across a 10-ohm resistor with 1 ampere of current equals 10 volts.
- The voltage across a 10-ohm resistor with 2 amperes of current equals 20 volts.

3.6 SUBELEMENT T6

Subelement T6 contains four targeted question groups with a total of 48 potential test questions. Four of the 35 [Element 2] test questions will be drawn from subelement T6.

Group T6A – Electrical Components

Questions in group T6A cover the following topics: fixed and variable resistors, capacitors, inductors, fuses, switches, and batteries.

T6A Test Notes

- Resistors are used to oppose the flow of current in a DC circuit.
- Potentiometers are often used as adjustable volume controls.
- Potentiometers control resistance.
- Capacitors store energy in an electric field.
- Capacitors consist of two or more conductive surfaces separated by an insulator.
- Inductors store energy in a magnetic field.
- Inductors are usually composed of coils of wire.
- Switches are used to connect or disconnect electrical circuits.
- Fuses are used to protect other circuit components from current overloads.
- Lead-acid gel-cell, nickel-metal hydride, and lithium-ion batteries are all rechargeable.
- Carbon-zinc batteries are not rechargeable.

Group T6B – Semiconductors

Questions in group T6B cover the basic principles and applications of common diodes and transistors.

T6B Test Notes

- Transistors are capable of using a voltage or current signal to control current flow.
- Diodes allow current to flow in only one direction.
- Transistors can be used as electronic switches or amplifiers.
- A transistor can be made of three layers of semiconductor material.

- Transistors can amplify signals.
- The cathode lead of a semiconductor diode is usually identified with a stripe.
- LED stands for light-emitting diode.
- FET stands for field-effect transistor.
- The two electrodes of a diode are anode and cathode.
- The three electrodes of a PNP or NPN transistor are emitter, base, and collector.
- The three electrodes of an FET are source, gate, and drain.
- Gain is a transistor's ability to amplify a signal.

Group T6C – Circuit Diagrams

Questions in group T6C center on recognizing basic schematic symbols used in circuit diagrams. Most questions from this group are based on one of schematic diagrams shown below.

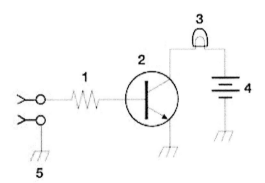

Figure T-1

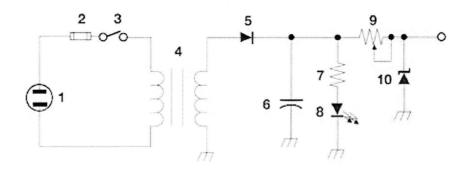

Figure T-2

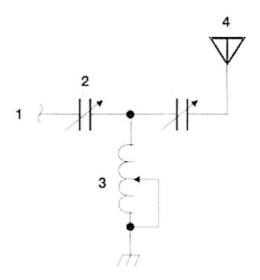

Figure T-3

T6C Test Notes

- Schematic symbols are standardized representations of components in an electrical wiring diagram.
- Component 1 in Figure T-1 represents a resistor.
- Component 2 in Figure T-1 represents a transistor.
- Component 3 in Figure T-1 represents a lamp.
- Component 4 in Figure T-1 represents a battery.
- Component 6 in Figure T-2 represents a capacitor.
- Component 8 in Figure T-2 represents a light-emitting diode (LED).

- Component 9 in Figure T-2 represents a variable resistor.
- Component 4 in Figure T-2 represents a transformer.
- Component 3 in Figure T-3 represents a variable inductor.
- Component 4 in Figure T-3 represents an antenna.
- The symbols on an electrical circuit schematic diagram represent electrical components.
- Electrical circuit schematic diagrams represent the way components are interconnected.

Group T6D – Component Functions

Questions in group T6D cover the following topics: rectification, switches, indicators, power supplies, resonant circuits, shielding, power transformers, and integrated circuits. Some questions may also refer to figures presented in the previous section (see question group T6C – Circuit Diagrams).

T6D Test Notes

- Rectifiers change an alternating current into a varying direct current signal.
- A relay is a switch controlled by an electromagnet.
- Component 3 in Figure T-2 represents a single-pole single-throw switch.
- A meter can be used to display signal strength on a numeric scale.
- A regulator controls the amount of voltage from a power supply.
- Transformers are used to change 120V AC house currents to lower AC voltages for other uses.
- An LED is commonly used as a visual indicator.
- A capacitor is used together with an inductor to make a tuned circuit.
- Integrated circuits combine several semiconductors and other components into one package.
- Component 2 in Figure T-1 controls the flow of current.
- A simple resonant or tuned circuit is an inductor and a capacitor connected in series or parallel to form a filter.

- A common reason to use shielded wire is to prevent coupling of unwanted signals to or from the wire.

3.7 SUBELEMENT T7

Subelement T7 contains four targeted question groups with a total of 48 potential test questions. Four of the 35 [Element 2] test questions will be drawn from subelement T7.

Group T7A – Station Equipment

Questions in group T7A cover the following topics: transmitters, receivers, transceivers, modulation, transverters, low power and weak signal operation, plus transmit and receive amplifiers.

T7A Test Notes

- Sensitivity is the ability of a receiver to detect the presence of a signal.
- A transceiver combines the functions of a transmitter and a receiver.
- Mixers are used to convert a radio signal from one frequency to another.
- Selectivity is the ability of a receiver to discriminate between multiple signals.
- Oscillators generate a signal of a desired frequency.
- A transverter takes the output of a low-powered 28 MHz SSB exciter and produces a 222 MHz output signal.
- The push-to-talk function (PTT) switches between receive and transmit.
- Modulation describes combining speech with an RF carrier signal.
- A multimode VHF transceiver is the most useful device for VHF weak-signal communication.
- RF power amplifiers increase the low-power output from a handheld transceiver.
- RF preamplifiers are installed between the antenna and receiver.

Group T7B – Transmitter and Receiver Problems

Questions in group T7B cover the following topics: overload and overdrive, distortion, interference and consumer electronics, Part 15 devices, over and under modulation, RF feedback, off frequency signals, fading and noise, and problems with digital communications interfaces.

T7B Test Notes

- Talk farther away from the microphone if you're FM handheld or mobile transceiver is over-deviating.
- A broadcast AM or FM radio could unintentionally receive an amateur radio transmission if the receiver is unable to reject strong signals outside the band.
- Radio frequency interference can be caused by fundamental overload, harmonics, and spurious emissions.
- One way to reduce or eliminate interference by an amateur transmitter to a nearby telephone is to put an RF filter on the telephone.
- Overload of a non-amateur radio or TV receiver by an amateur signal can be reduced or eliminated by blocking the amateur signal with a filter at the antenna input of the affected receiver.
- If a neighbor tells you that your station transmissions are interfering with their radio or TV reception, make sure that your station is functioning properly and that it does not cause interference to your own radio or television when it is tuned to the same channel.
- Ferrite chokes, low-pass and high-pass filters, plus band-reject and band-pass filters are all useful in correcting radio frequency interference problems.
- If something in a neighbor's home is causing harmful interference to your amateur station, check your station and make sure it meets the standards of good amateur practice, work with your neighbor to identify the offending device, and politely inform your neighbor about the rules that prohibit the use of devices that cause interference.

- A Part 15 device is an unlicensed device that may emit low-powered radio signals on frequencies used by a licensed service.
- If you're audio signal through the repeater is distorted or unintelligible, your transmitter may be slightly off frequency, your batteries may be running low, or you could be in a bad location.
- Garbled, distorted, or unintelligible transmissions are symptoms of RF feedback in a transmitter or transceiver.
- A first step to resolve cable TV interference from your ham radio transmission is to be sure all TV coaxial connectors are installed properly.

Group T7C – Antenna Measurements and Troubleshooting

Questions in group T7C cover the following topics: measuring SWR, dummy loads, coaxial cables, and feed line failure modes.

T7C Test Notes

- The primary purpose of a dummy load is to prevent the radiation of signals during testing.
- An antenna analyzer can be used to determine if an antenna is resonant at the desired operating frequency.
- Standing-wave ratio (SWR) is a measure of how well a load is matched to a transmission line.
- An SWR meter reading of 1 to 1 (1:1) indicates a perfect impedance match between the antenna and feed line.
- The approximate SWR value above which the protection circuits in most solid-state transmitters begin to reduce transmitter power is 2 to 1 (2:1).
- An SWR reading of 4:1 indicates impedance mismatch.
- Power lost in a feed line is converted into heat.
- A directional wattmeter could also be used to determine if a feed line and antenna are properly matched.
- The most common cause for failure of coaxial cables is moisture contamination.
- The outer jacket of coaxial cable should be resistant to ultraviolet light because it can damage the jacket and allow water to enter the cable.

- A disadvantage of air core coaxial cable is that it requires special techniques to prevent water absorption.
- A common use of coaxial cable is carrying RF signals between a radio and antenna.
- A dummy load consists of a non-inductive resistor and a heat sink.

Group T7D – Basic Testing and Repair

Questions in group T7D cover the use of basic test instruments and soldering.

T7D Test Notes

- A voltmeter measures electric potential or electromotive force.
- The correct way to connect a voltmeter is in parallel to a circuit.
- An ammeter is usually connected in series with a circuit.
- An ammeter is used to measure electric current.
- An ohmmeter is used to measure resistance.
- Attempting to measure voltage when using the resistance setting might damage a multimeter.
- Voltage and resistance measurements are commonly made using a multimeter.
- Rosin-core solder is best for radio and electronic use.
- The surface of a cold solder joint appears grainy or dull.
- When an ohmmeter connected across an unpowered circuit initially indicates a low resistance and then shows increasing resistance with time, the circuit probably contains a large capacitor.
- Ensure that the circuit is not powered when measuring resistance with an ohmmeter.
- When measuring high voltages, ensure that the voltmeter and leads are rated for use at the voltages to be measured.

3.8 SUBELEMENT T8

Subelement T8 contains four targeted question groups with a total of 45 potential test questions. Four of the 35 [Element 2] test questions will be drawn from subelement T8.

Group T8A – Modulation Modes

Questions in group T8A cover signal bandwidths and choice of emission type.

T8A Test Notes

- Single sideband (SSB) is a form of amplitude modulation (AM).
- Frequency modulation (FM) is most commonly used for VHF packet radio transmissions.
- SSB is most often used for long-distance (weak signal) contacts on VHF and UHF bands.
- FM is most commonly used for VHF and UHF voice repeaters.
- Continuous wave emission (CW) has the narrowest bandwidth.
- Upper sideband (USB) is normally used for 10 meter HF, VHF and UHF single-sideband communications.
- The primary advantage of single sideband over FM for voice transmissions is that SSB signals have narrower bandwidth.
- The approximate bandwidth of a single-sideband voice signal is 3 kHz.
- The approximate bandwidth of a VHF repeater FM phone signal is between 10 and 15 kHz.
- The typical bandwidth of analog fast-scan TV transmissions on the 70 cm band is about 6 MHz.
- The approximate maximum bandwidth required to transmit a CW signal is 150 Hz.

Group T8B – Amateur Satellite Operation

Questions in group T8B cover the following topics: Doppler shift, basic orbits, operating protocols, control operator, transmitter power considerations, satellite tracking, and digital modes.

T8B Test Notes

- Any amateur whose license privileges allow him or her to transmit on the satellite uplink frequency may be the control operator of a station communicating through an amateur satellite or space station.
- Only the minimum amount of power needed to complete the contact should be used on the uplink frequency of an amateur satellite or space station.
- Satellite tracking programs provide maps showing the real-time position of the satellite track over the earth; the time, azimuth, and elevation of the start, maximum altitude and end of a pass; plus the apparent frequency of the satellite transmission with effects of Doppler shift.
- Any amateur holding a Technician or higher class license may make contact with an amateur station on the International Space Station using 2 meter and 70 cm band amateur radio frequencies.
- A satellite beacon is a transmission from a space station that contains information about a satellite.
- Keplerian elements are inputs to a satellite tracking program.
- With regard to satellite communications, Doppler shift is an observed change in signal frequency caused by relative motion between the satellite and the earth station.
- A satellite is operating in mode U/V when the satellite uplink is in the 70 cm band and the downlink is in the 2 meter band.
- Spin fading of a satellite signal is caused by rotation of the satellite and its antennas.
- The initials LEO tell you that a satellite is in a low earth orbit.
- FM Packet is a commonly used method of sending signals to and from a digital satellite.

Group T8C – Operating Activities

Questions in group T8C cover the following topics: radio direction finding, radio control, contests, linking over the Internet, and grid locators.

T8C Test Notes

- Radio direction finding is used to locate sources of noise interference or jamming.
- A directional antenna is useful for hidden transmitter hunts.
- Contesting involves contacting as many stations as possible during a specified period of time.
- Good procedure when contacting another station in a radio contest is sending only the minimum information needed for proper identification and the contest exchange.
- A grid locator is a letter-number designator assigned to a geographic location.
- An internet radio linking project (IRLP) node is accomplished by using dual-tone multi-frequency (DTMF) signals.
- The maximum power allowed when transmitting telecommand signals to radio-controlled models is 1 watt.
- A label indicating the licensee's name, call sign, and address must be affixed to the transmitter in place of on-air station identification when sending signals to a radio-controlled model using amateur frequencies.
- You can obtain a list of active nodes that use VoIP from a repeater directory.
- Use the keypad to transmit a specific IRLP node ID when using a portable transceiver.
- A gateway is an amateur radio station that is used to connect other amateur stations to the Internet.
- In amateur radio, Voice Over Internet Protocol (VoIP) is a method of delivering voice communications over the Internet using digital techniques.
- The Internet Radio Linking Project (IRLP) is a technique to connect amateur radio systems, such as repeaters, via the Internet using VoIP.

Group T8D – Non-voice Communications

Questions in group T8D cover the following topics: image signals, digital modes, CW, packet, PSK31, APRS, error detection and correction, and NTSC.

T8D Test Notes

- Packet, PSK31, and MFSK are all forms of digital communications.
- The abbreviation APRS stands for Automatic Packet Reporting System.
- A Global Positioning System receiver provides data to the transmitter when sending automatic position reports from a mobile amateur radio station.
- The term NTSC refers to an analog fast-scan color TV signal.
- Providing real-time tactical digital communications in conjunction with a map showing the locations of stations is an application of APRS.
- The abbreviation PSK stands for phase shift keying.
- PSK31 is a low-rate data transmission mode.
- A packet transmission can include a header which contains the call sign of the station to which the information is being sent, a check sum which permits error detection, and an automatic repeat request in case of error.
- International Morse code is used when sending CW in the amateur bands.
- A straight key, electronic keyer, or computer keyboard can all be used to transmit CW in the amateur bands.
- An ARQ transmission system is a digital scheme whereby the receiving station detects errors and sends a request to the sending station to retransmit the information.

3.9 SUBELEMENT T9

Subelement T9 contains two targeted question groups with a total of 25 potential test questions. Two of the 35 [Element 2] test questions will be drawn from subelement T9.

Group T9A – Antennas

Questions in group T9A cover the following topics: vertical and horizontal polarization, gain; common portable and mobile antennas, and the relationships between antenna length and frequency.

T9A Test Notes

- A beam antenna is an antenna that concentrates signals in one direction.
- The electric field of a vertical antenna is perpendicular to the Earth.
- A horizontally polarized antenna is a simple dipole mounted so the conductor is parallel to the Earth's surface.
- The "rubber duck" antenna supplied with most handheld radio transceivers does not transmit or receive as effectively as a full-sized antenna.
- You can shorten a dipole antenna to make it resonant on a higher frequency.
- Quad, Yagi, and dish antennas are directional antennas.
- A good reason not to use a "rubber duck" antenna inside your car is because signals can be significantly weaker than when it is outside of the vehicle.
- The approximate length of a 1/4-wavelength vertical antenna for 146 MHz is 19 inches.
- The approximate length of a 6 meter 1/2-wavelength wire dipole antenna is 112 inches.
- Radiation in free space is strongest from a 1/2-wavelength dipole antenna in the broadside direction to the antenna.
- The gain of an antenna is the increase in signal strength in a specified direction when compared to a reference antenna.

- Using a properly mounted 5/8 wavelength antenna for VHF or UHF mobile service offers a lower angle of radiation and more gain than a 1/4-wavelength antenna and usually provides improved coverage.
- VHF or UHF mobile antennas are often mounted in the center of the vehicle roof because a roof-mounted antenna normally provides the most uniform radiation pattern.
- Inserting an inductor in the radiating portion of the antenna to make it electrically longer is an antenna loading technique.

Group T9B – Feed Lines

Questions in group T9B cover the following topics: types of feed lines, attenuation vs. frequency, SWR concepts, matching, weather protection, plus choosing RF connectors and feed lines.

T9B Test Notes

- Low SWR in an antenna system that uses a coaxial cable feed line allows efficient transfer of power and reduces loss.
- The most commonly used coaxial cable in typical amateur radio installations has an impedance of 50 ohms.
- Coaxial cable is used more often than any other feed line for amateur radio antenna systems because it is easy to use and requires few special installation considerations.
- An antenna tuner matches the antenna system impedance to the transceiver's output impedance.
- Loss generally increases as the frequency of a signal passing through coaxial cable is increased.
- A Type N connector is most suitable for frequencies above 400 MHz.
- PL-259 type coax connectors are commonly used at HF frequencies.
- Coax connectors exposed to the weather should be sealed against water intrusion to prevent an increase in feed line loss.
- A loose connection in an antenna or a feed line might cause erratic changes in SWR.

- RG-8 cable is larger in size and has less loss at a given frequency than RG-58.
- Air-insulated hard line has the lowest loss at VHF and UHF.

3.10 SUBELEMENT T0

Subelement T0 contains three targeted question groups with a total of 36 potential test questions. Three of the 35 [Element 2] test questions will be drawn from subelement T0.

Group T0A – Power Circuits and Hazards

Questions in group T0A cover the following topics: hazardous voltages, fuses and circuit breakers, grounding, lightning protection, battery safety, and electrical code compliance.

T0A Test Notes

- Shorting the terminals of a 12-volt storage battery can cause burns, fire, or an explosion.
- Current flowing through the body causes a health hazard by heating tissue, disrupting the electrical functions of cells, and causing involuntary muscle contractions.
- The green wire in a three-wire electrical AC plug is connected to safety ground.
- The purpose of a fuse in an electrical circuit is to interrupt power in case of overload.
- It is unwise to install a 20-ampere fuse in the place of a 5-ampere fuse because excessive current could cause a fire.
- Guard against electrical shock at your station by using three-wire cords and plugs for all AC-powered equipment, connecting all AC-powered station equipment to a common safety ground, and using a circuit protected by a ground-fault interrupter.

- When installing devices for lightning protection in a coaxial cable feed line, ground all of the protectors to a common plate that is connected to an external ground.
- A fuse or circuit breaker in series with the AC hot conductor should always be included in home-built equipment that is powered from 120 V AC power circuits.
- Explosive gas can collect if 12-volt storage batteries are not properly vented.
- A lead-acid storage battery could overheat and give off flammable gas or explode if charged or discharged too quickly.
- You can receive an electric shock from the charge stored in large capacitors inside power supplies, even when they are turned off and disconnected.

Group T0B – Antenna Safety

Questions in group T0B cover the following topics: tower safety, antenna supports, overhead power lines, and installation.

T0B Test Notes

- Members of a tower work team should wear a hard hat and safety glasses at all times when any work is being done on the tower.
- Put on a climbing harness and safety glasses before climbing an antenna tower.
- It is NEVER safe to climb a tower without a helper or an observer.
- Look for and stay clear of any overhead electrical wires when putting up an antenna tower.
- A gin pole is used to lift tower sections or antennas.
- When installing an antenna, allow for a minimum safe distance, so that if the antenna falls unexpectedly during installation, no part of it can come closer than 10 feet to any power lines.
- Crank-up towers must never be climbed unless in the fully retracted position.

- To properly ground a tower, use separate eight-foot long ground rods that are bonded to the tower and each other for each tower leg.
- Avoid attaching an antenna to a utility pole, because the antenna could come in contact with high-voltage power wires.
- Sharp bends must be avoided in grounding conductors used for lightning protection.
- Local electrical codes establish the grounding requirements for an amateur radio tower or antenna.
- Ensure that connections are short and direct when installing ground wires on a tower for lightning protection.

Group T0C – RF Hazards

Questions in group T0C cover the following topics: radiation exposure, proximity to antennas, recognized safe power levels, exposure to others, radiation types, and duty cycle.

T0C Test Notes

- VHF and UHF radio signals produce non-ionizing radiation.
- 50 MHz has the lowest value for maximum permissible exposure (MPE) limit.
- The maximum power level that an amateur radio station may use at VHF frequencies before a required RF exposure evaluation is 50 watts peak envelope power (PEP) at the antenna.
- Frequency and power level of the RF field, distance from the antenna to a person, and radiation pattern of the antenna all affect the RF exposure of people near an amateur station antenna.
- Exposure limits vary with frequency because the human body absorbs more RF energy at some frequencies than at others.
- Calculation based on FCC OET Bulletin 65, calculation based on computer modeling, and measurement of field strength using calibrated equipment are all acceptable methods to determine if a station complies with FCC RF exposure regulations.

- If a person accidentally touched your antenna while you were transmitting, they might receive a painful RF burn.
- Amateur operators can relocate antennas to prevent exposure to RF radiation in excess of FCC-supplied limits.
- Make sure your station stays in compliance with RF safety regulations by reevaluating the station whenever an item of equipment is changed.
- Duty cycle is one of the factors used to determine safe RF radiation exposure levels because it affects the average exposure of people to radiation.
- Duty cycle during the averaging time for RF exposure is the percentage of time that a transmitter is transmitting.
- RF radiation differs from ionizing radiation (radioactivity) because RF radiation does not have sufficient energy to cause genetic damage.
- If the averaging time for exposure is 6 minutes, two (2) times as much power density is permitted if the signal is present for 3 minutes and absent for 3 minutes rather than being present for the entire 6 minutes.

3.11 STUDY TIPS

Review the test notes for each subelement carefully. Once you are comfortable with the majority of information presented, take the practice exam in Appendix 1 to check your progress. Repeat the process as needed.

You should also try a random sampling of the Element 2 question pool for more practice. The resource directory in Appendix 4 will show you not only where to download the entire question pool online but also where to find detailed explanations for all material covered on the licensing exam. Everything you need to pass your Amateur Radio (Ham) License Examination is in this manual – all you have to do is use the information, study with diligence, and practice!

4. GENERAL CLASS (ELEMENT 3)

General Class written examinations are referred to as Element 3. Topics covered during examination include: FCC rules, operating procedures, radio wave propagation, amateur radio practices, electrical principles, circuit components, practical circuits, signals and emissions, antennas and feed lines, plus electrical and RF safety.

Element 3 exams are derived from 10 subelements containing 35 targeted question groups. The current public question pool contains a total of 456 available test questions. Element 3 tests are made up of 35 random questions, of which at least 26 must be answered correctly to qualify for a General Class operator license. Following is a rundown of applicable subelements and related question groups.

4.1 SUBELEMENT G1

Subelement G1 contains five targeted question groups with a total of 58 potential test questions. Five of the 35 [Element 3] test questions will be drawn from subelement G1.

Group G1A – FCC Rules

Questions in group G1A cover the following topics: General Class control operator frequency privileges, plus primary and secondary allocations.

G1A Test Notes

- A General Class license holder is granted all amateur frequency privileges on the 160, 60, 30, 17, 12, and 10 meter bands.
- Phone operation is prohibited on the 30 meter band.
- Image transmission is prohibited on the 30 meter band.
- The 60 meter amateur band is restricted to communication on only specific channels rather than frequency ranges.
- 7.250 MHz is within the General Class portion of the 40 meter band.
- 24.940 MHz is within the 12 meter band.
- 3900 kHz is within the General Class portion of the 75 meter phone band.
- 14305 kHz is within the General Class portion of the 20 meter phone band.
- 3560 kHz is within the General Class portion of the 80 meter band.
- 21300 kHz is within the General Class portion of the 15 meter band.
- The frequencies 28.020 MHz, 28.350 MHz, and 28.550 MHz are all available to a control operator holding a General Class license.
- When General Class licensees are not permitted to use the entire voice portion of a particular band, the upper frequency portion of the voice segment is generally available to them.
- No amateur service band is shared with the Citizens Radio Service.
- When the FCC designates the Amateur Service as a secondary user on a band, amateur stations are allowed to use the band only if they do not cause harmful interference to primary users.

- If a station in the primary service interferes with your contact when operating on either the 30 or 60 meter bands, move to a clear frequency.

Group G1B – FCC Rules

Questions in group G1B cover the following topics: antenna structure limitations, good engineering and amateur practice, beacon operation, restricted operation, and retransmitting radio signals.

G1B Test Notes

- An antenna structure may be erected at a maximum height of 200 feet above ground without requiring notification to the FAA and registration with the FCC, provided it is not at or near a public-use airport.
- There must be no more than one beacon signal in the same band from a single location.
- One purpose of a beacon station is for observation of propagation and reception.
- Before amateur stations may provide communications to broadcasters for dissemination to the public, the communications must directly relate to the immediate safety of human life or protection of property and there must be no other means of communication reasonably available before or at the time of the event.
- An amateur station may transmit music when it is an incidental part of a manned spacecraft retransmission.
- An amateur station is permitted to transmit secret codes only to control a space station.
- Abbreviations or procedural signals may be used in the Amateur Service if they do not obscure the meaning of a message.
- When choosing a transmitting frequency, review the FCC Part 97 rules regarding permitted frequencies and emissions, follow generally accepted band plans agreed to by the amateur radio community, and listen to avoid interfering with ongoing communication before transmitting.
- An amateur station may transmit communications in which the licensee or control operator has a pecuniary (monetary) interest when other amateurs

are being notified of the sale of apparatus normally used in an amateur station, and such activity is not done on a regular basis.
- The power limit for a beacon station is 100 watts PEP output.
- In all respects not specifically covered by the Part 97 rules, the FCC requires amateur stations to be operated in conformance with good engineering and good amateur practice.
- The FCC determines "good engineering and good amateur practice" as applied to the operation of an amateur station in all respects not covered by the Part 97 rules.

Group G1C – FCC Rules

Questions in group G1C cover the following topics: transmitter power regulations and data emission standards.

G1C Test Notes

- The maximum transmitting power an amateur station may use on 10.140 MHz is 200 watts PEP output.
- The maximum transmitting power an amateur station may use on the 12 meter band is 1500 watts PEP output.
- The maximum bandwidth permitted for Amateur Radio stations when transmitting on USB frequencies in the 60 meter band is 2.8 kHz.
- Only the minimum power necessary to carry out the desired communications should be used on the 14 MHz band.
- Transmitter power is limited to 1500 watts PEP output on the 28 MHz band.
- Transmitter power is limited to 1500 watts PEP output on the 1.8 MHz band.
- The maximum symbol rate permitted for radioteletype (RTTY) or data emission transmission on the 20 meter band is 300 baud.
- The maximum symbol rate permitted for RTTY or data emission transmitted at frequencies below 28 MHz is 300 baud.
- The maximum symbol rate permitted for RTTY or data emission transmitted on the 1.25 meter and 70 centimeter bands is 56 kilobaud.

- The maximum symbol rate permitted for RTTY or data emission transmissions on the 10 meter band is 1200 baud.
- The maximum symbol rate permitted for RTTY or data emission transmissions on the 2 meter band is 19.6 kilobaud.

Group G1D – FCC Rules

Questions in group G1D cover the following topics: Volunteer Examiners (VEs), Volunteer Examiner Coordinators (VECs), and temporary identification.

G1D Test Notes

- A proper way to identify when transmitting using phone on General Class frequencies if you have a Certificate of Successful Completion of Examination (CSCE) is to give your call sign followed by "slant AG".
- You may only administer a Technician license examination when you are an accredited VE holding a General Class operator license.
- You may operate on any General or Technician Class band segment if you are a Technician Class operator and have a CSCE for General Class privileges.
- At least three VEC-accredited General Class or higher VEs must be present for administering a Technician Class operator examination.
- An FCC General Class or higher license and VEC accreditation is sufficient for you to be an administering VE for a Technician Class operator license examination.
- If you are a Technician Class licensee with a CSCE for General Class operator privileges but the FCC has not yet posted your upgrade on its Web site, add the special identifier "AG" after your call sign whenever you operate using General Class privileges.
- Volunteer Examiners are accredited by a Volunteer Examiner Coordinator.
- For a non-U.S. citizen to be an accredited Volunteer Examiner, the person must hold an FCC-granted Amateur Radio license of General Class or above.

- A Certificate of Successful Completion of Examination (CSCE) is valid for 365 days.
- The minimum age to qualify as an accredited Volunteer Examiner is 18 years old.

Group G1E – FCC Rules

Questions in group G1E cover the following topics: control categories, repeater regulations, harmful interference, third-party rules, and ITU regions.

G1E Test Notes

- A third party is disqualified from participating in stating a message over an amateur station if their amateur license has ever been revoked.
- A 10 meter repeater may retransmit the 2 meter signal from a station with a Technician Class control operator only if the 10 meter repeater control operator holds at least a General Class license.
- ITU region 2 control operators holding a General Class license are permitted to operate in the 7.175 to 7.300 MHz band.
- An Amateur Radio station licensee is required to take specific steps in avoiding harmful interference to other users or facilities when operating within one mile of an FCC monitoring station, when using a band where the Amateur Service is secondary, or when a station is transmitting spread spectrum emissions.
- Only messages relating to Amateur Radio, remarks of a personal character, or messages relating to emergencies or disaster relief may be transmitted by an amateur station for a third party in another country.
- In the event of interference between a coordinated repeater and an uncoordinated repeater, the licensee of the non-coordinated repeater has primary responsibility to resolve the interference.
- Third-party traffic, except for messages directly involving emergencies or disaster relief communications, is prohibited in every foreign country unless there is a third-party agreement in effect with that country.
- For a non-licensed person to communicate with a foreign Amateur Radio station from a station with an FCC-granted license at which a licensed

control operator is present, the foreign amateur station must be in a country with which the United States has a third-party agreement.
- If using a language other than English in making a contact using phone emission, you must use English when identifying your station.
- The portion of the 10 meter band above 29.5 MHz is available for repeater use.

4.2 SUBELEMENT G2

Subelement G2 contains five targeted question groups with a total of 58 potential test questions. Five of the 35 [Element 3] test questions will be drawn from subelement G2.

Group G2A – Operating Procedures

Questions in group G2A cover the following topics: phone operating procedures, USB/LSB utilization conventions, procedural signals, breaking into a QSO in progress, and VOX operation.

G2A Test Notes

- Upper sideband is most commonly used for voice communications on frequencies of 14 MHz or higher.
- Lower sideband is most commonly used for voice communications on the 160, 75, and 40 meter bands.
- Upper sideband is most commonly used for SSB voice communications in the VHF and UHF bands.
- Upper sideband is most commonly used for voice communications on the 17 and 12 meter bands.
- Single-sideband voice communication most commonly used on the high frequency amateur bands.
- The advantages of using a single sideband as compared to other analog voice modes on the HF amateur bands are less bandwidth usage and higher power efficiency.
- In a single-sideband voice mode, only one sideband is transmitted while the other sideband and carrier are suppressed.
- Break into a conversation when using phone by saying your call sign during a break between transmissions from the other stations.
- Current amateur practice is to use lower sideband on the 160, 75, and 40 meter bands.
- SSB VOX operation allows for "hands-free" operations.
- "CQ DX" usually indicates that the caller is looking for any station outside his or her own country.

Group G2B – Operating Procedures

Questions in group G2B cover the following topics: operating courtesy, band plans, emergencies, plus drills and emergency communications.

G2B Test Notes

- No one has priority access to frequencies, and common courtesy should be a guide.
- The first thing you should do if you are communicating with another amateur station and hear a station in distress break in is to acknowledge the station in distress and determine what assistance may be needed.
- If propagation changes during your contact and you notice increasing interference from other activity on the same frequency, move your contact to another frequency as a common courtesy.
- When selecting a CW transmitting frequency, you should allow a minimum frequency separation of 150 to 500 Hz in order to minimize interference to stations on adjacent frequencies.
- The customary minimum frequency separation between SSB signals under normal conditions is approximately 3 kHz.
- A practical way to avoid harmful interference is to send "QRL?" on CW, followed by your call sign; or if using phone, ask if the frequency is in use, followed by your call sign.
- To comply with good amateur practice when choosing a frequency on which to initiate a call, follow the voluntary band plan for the operating mode you intend to use.
- The "DX window" in a voluntary band plan is a portion of the band that should not be used for contacts between stations within the 48 contiguous United States.
- Only a person holding an FCC-issued amateur operator license may be the control operator of an amateur station transmitting in RACES to assist in relief operations during a disaster.
- The FCC may restrict normal frequency operations of amateur stations participating in RACES when the President's War Emergency Powers have been invoked.
- To send a distress call, use whatever frequency has the best chance of communicating the distress message.

- An amateur station is allowed to use any means at its disposal to assist another station in distress at any time during an actual emergency.

Group G2C – Operating Procedures

Questions in group G2C cover the following topics: CW operating procedures and procedural signals, Q signals and common abbreviations, plus full break in protocol.

G2C Test Notes

- A full break-in telegraphy (QSK) is when transmitting stations can receive between code characters and elements.
- The CW message "QRS" is a request to send slower.
- The CW message "KN" at the end of a transmission means listening only for a specific station or stations.
- The CW message "CL" at the end of a transmission means the operator is closing the station.
- The speed at which the CQ was sent is the best speed to use answering a CQ in Morse code.
- "Zero beat" means matching your transmit frequency to the frequency of a received signal.
- When sending CW, a "C" added to the RST report means a chirpy or an unstable signal.
- "AR" is a prosign sent to indicate the end of a formal message when using CW.
- The Q signal "QSL" means "I acknowledge receipt."
- The Q signal "QRQ" means "send faster."
- The Q signal "QRV" means "I am ready to receive messages."

Group G2D – Operating Procedures

Questions in group G2D cover the following topics: Amateur Auxiliary, minimizing interference, and HF operations.

G2D Test Notes

- The Amateur Auxiliary is an organization of amateur volunteers who are formally enlisted to monitor the airwaves for rule violations.
- Objectives of the Amateur Auxiliary are encouraging amateur self-regulation and compliance with the rules.
- "Hidden transmitter hunts" help to develop direction-finding skills that can be used to locate stations violating FCC Rules.
- An azimuthal projection map is a world map projection centered on a particular location.
- It is permissible to communicate with amateur stations in countries outside the areas administered by the Federal Communications Commission when the contact is with amateurs in any country except those whose administrations have notified the ITU that they object to such communications.
- A directional antenna is pointed 180 degrees from its short-path heading when making "long-path" contact with another station.
- You must keep a record of the gain of your antenna when operating in the 60 meter band and using other than a dipole antenna.
- Although not required, many amateurs keep a log to help with a reply if the FCC requests information.
- Station logs typically record the date and time of contact, band and/or frequency of the contact, plus the contacted station's call sign and signal report given.
- QRP operation is a low-power transmit operation.
- A unidirectional antenna is the best HF antenna to use for minimizing interference.

Group G2E – Digital Operation

Questions in group G2E cover the following topics: procedures, procedural signals, and common abbreviations.

G2E Test Notes

- Lower-sideband (LSB) mode is normally used when sending an RTTY signal via audio frequency shift keying (AFSK) with a single-sideband (SSB) transmitter.
- The number of data bits that are sent in a single PSK31 character varies.
- The header part of a data packet contains routing and handling information.
- The 14.070 – 14.100 MHz segment of the 20 meter band is most often used for data transmissions.
- Baudot code is a 5-bit code with additional start and stop bits.
- 170 Hz is the most common frequency shift for RTTY emissions in the amateur HF bands.
- RTTY stands for radioteletype.
- The 3570 – 3600 kHz segment of the 80 meter band is most commonly used for data transmissions.
- Most PSK31 operations are found below the RTTY segment of the 20 meter band, near 14.070 MHz.
- A major advantage of MFSK16 is that it offers good performance in weak signal environments without error correction.
- MFSK stands for multi (or multiple) frequency shift keying.
- Receiving stations respond to an ARQ data mode packet containing errors by requesting the packet be retransmitted.
- In the PACTOR protocol, a NAK response is a request that the packet be retransmitted.

4.3 SUBELEMENT G3

Subelement G3 contains three targeted question groups with a total of 41 potential test questions. Three of the 35 [Element 3] test questions will be drawn from subelement G3.

Group G3A – Radio Wave Propagation

Questions in group G3A cover the following topics: sunspots and solar radiation, ionospheric disturbances, and propagation forecasting and indices.

G3A Test Notes

- The sunspot number is a measure of solar activity based on counting sunspots and sunspot groups.
- Sudden ionospheric disturbance disrupts daytime ionospheric propagation of HF radio waves on lower frequencies more than those on higher frequencies.
- It takes 8 minutes for the increased ultraviolet and X-ray radiation from solar flares to affect radio wave propagation on the Earth.
- 21 MHz and higher are the least reliable HF frequencies for long-distance communications during periods of low solar activity.
- The solar-flux index is a measure of solar radiation at 10.7 cm.
- A geomagnetic storm is a temporary disturbance in the Earth's magnetosphere.
- The 20 meter band usually supports worldwide propagation during daylight hours at any point in the solar cycle.
- A geomagnetic storm can degrade high-latitude HF radio wave propagation.
- High sunspot numbers enhance long-distance communications in the upper HF and lower VHF range.
- The sun's rotation on its axis causes HF propagation conditions to vary periodically in a 28-day cycle.
- The typical sunspot cycle lasts for approximately 11 years.
- The K-index indicates the short-term stability of Earth's magnetic field.

- The A-index indicates the long-term stability of Earth's geomagnetic field.
- HF communications are disturbed by the charged particles that reach the Earth from solar coronal holes.
- It takes 20 to 40 hours for charged particles from coronal mass ejections to affect radio wave propagation on the Earth.
- Periods of high geomagnetic activity may result in an aurora that can reflect VHF signals.

Group G3B – Radio Wave Propagation

Questions in group G3B cover the following topics: maximum usable frequency, lowest usable frequency, and propagation.

G3B Test Notes

- A sky-wave signal might sound like a well-defined echo if it arrives at your receiver by both short-path and long-path propagation.
- Short skip sky-wave propagation on the 10 meter band is a good indicator of the possibility of sky-wave propagation on the 6 meter band.
- When choosing a transmit frequency on HF, select a frequency just below the maximum usable frequency (MUF) for lowest attenuation.
- To determine if the MUF is high enough to support skip propagation between your station and a distant location on frequencies between 14 and 30 MHz, listen for signals from an international beacon.
- When radio waves with frequencies below the MUF and above the lowest usable frequency (LUF) are sent into the ionosphere, they are usually bent back to the Earth.
- Radio waves with frequencies below the LUF are usually completely absorbed by the ionosphere.
- LUF stands for the lowest usable frequency for communications between two points.
- MUF stands for the maximum usable frequency for communications between two points.
- The maximum distance along the Earth's surface normally covered in one hop using the F2 region is approximately 2,500 miles.

- The maximum distance along the Earth's surface normally covered in one hop using the E region is approximately 1,200 miles.
- When the LUF exceeds the MUF, no HF radio frequency will support ordinary sky-wave communications over the path.
- Path distance and location, time of day and season, plus solar radiation and ionospheric disturbances all affect the MUF.

Group G3C – Radio Wave Propagation

Questions in group G3C cover the following topics: Ionospheric layers, critical angle and frequency, HF scatter, plus near vertical incidence sky-waves.

G3C Test Notes
- The D layer is the closest ionospheric layer to Earth's surface.
- Where the sun is overhead are the spots on Earth that ionospheric layers reach their maximum height.
- The F2 region is mainly responsible for the longest distance radio wave propagation because it is the highest ionospheric region.
- "Critical angle" means the highest take-off angle that will return a radio wave to the Earth under specific ionospheric conditions.
- Long-distance communication on the 40, 60, 80 and 160 meter bands are more difficult during the day because the D layer absorbs signals at these frequencies during daylight hours.
- HF scatter signals have a wavering sound.
- HF scatter signals often sound distorted because energy is scattered into the skip zone through several different radio wave paths.
- HF scatter signals in the skip zone are usually weak because only a small part of the signal energy is scattered into the skip zone.
- Scatter propagation allows a signal to be detected at a distance too far for ground-wave propagation but too near for normal sky-wave propagation.
- An HF signal heard on a frequency above the MUF is an indication that the signal is being received via scatter propagation.

- Horizontal dipoles placed between 1/8 and 1/4 wavelength above ground will be the most effective antenna type for skip communications on 40 meters during the day.
- The D layer is the most absorbent ionospheric layer of long skip signals during daylight hours on frequencies below 10 MHz.
- Near vertical incidence sky-wave (NVIS) propagation is short distance HF propagation using high elevation angles.

4.4 SUBELEMENT G4

Subelement G4 contains five targeted question groups with a total of 65 potential test questions. Five of the 35 [Element 3] test questions will be drawn from subelement G4.

Group G4A – Amateur Radio Practices

Questions in group G4A cover station operation and setup.

G4A Test Notes

- "Notch filters" are for reducing interference from carriers in the receiver passband.
- One advantage of selecting the opposite or "reverse" sideband when receiving CW signals on a typical HF transceiver is that it may be possible to reduce or eliminate interference from other signals.
- Operating in "split" mode normally means the transceiver is set to different transmit and receive frequencies.
- A pronounced dip on the plate current meter of a vacuum tube RF power amplifier indicates correct adjustment of the plate tuning control.
- The purpose of using automatic level control (ALC) with an RF power amplifier is to reduce distortion due to excessive drive.
- An antenna coupler is often used to enable matching the transmitter output to an impedance other than 50 ohms.

- Excessive drive power can lead to permanent damage when using a solid-state RF power amplifier.
- The correct adjustment for the load or coupling control of a vacuum tube RF power amplifier is maximum power output without exceeding maximum allowable plate current.
- A time delay is sometimes included in a transmitter keying circuit to allow transmit-receive changeover operations to complete properly before RF output is allowed.
- The purpose of an electronic keyer is for automatic generation of strings of dots and dashes for CW operation.
- The intermediate frequency (IF) shift control on a receiver is used to avoid interference from stations very close to the receive frequency.
- The dual VFO feature on a transceiver permits ease of monitoring of the transmit and receive frequencies when they are not the same.
- One reason to use the attenuator function that is present on many HF transceivers is to reduce signal overload due to strong incoming signals.
- When transmitting PSK31 data signals, the transceiver audio input should be adjusted so that the transceiver ALC system does not activate.

Group G4B – Amateur Radio Practices

Questions in group G4B cover test and monitoring equipment and two-tone testing.

G4B Test Notes

- An oscilloscope contains horizontal and vertical channel amplifiers.
- An advantage of an oscilloscope over a digital voltmeter is that an oscilloscope can be used to measure complex waveforms.
- An oscilloscope is the best instrument to use when checking the keying waveform of a CW transmitter.
- The attenuated RF output of the transmitter is connected to the vertical input of an oscilloscope when checking the RF envelope pattern of a transmitted signal.

- High input impedance is desirable for a voltmeter because it decreases the loading on circuits being measured.
- A digital voltmeter has better precision than an analog voltmeter for most uses.
- A field-strength meter can be used for close-in radio direction-finding.
- A field-strength meter can also be used to monitor relative RF output when making antenna and transmitter adjustments.
- The radiation pattern of an antenna can be determined with a field-strength meter.
- Standing-wave ratio can be determined with a directional wattmeter.
- An antenna and feed line must be connected to an antenna analyzer when it is being used for SWR measurements.
- Strong signals from nearby transmitters can affect the accuracy when making measurements on an antenna system with an antenna analyzer.
- An antenna analyzer can be used for measuring the SWR of an antenna system and determining the impedance of an unknown or unmarked coaxial cable.
- Use of an instrument with an analog readout may be preferred over an instrument with a digital readout when adjusting tuned circuits.
- A two-tone test analyzes transmitter linearity.
- Two non-harmonically-related audio signals are used to conduct a two-tone test.

Group G4C – Amateur Radio Practices

Questions in group G4C cover the following topics: Interference with consumer electronics, grounding, and digital signal processing (DSP).

G4C Test Notes

- A bypass capacitor may be useful in reducing RF interference to audio-frequency devices.
- Interference over a wide range of frequencies could be caused by arcing at a poor electrical connection.

- Interference from a nearby single-sideband phone transmitter can produce distorted speech in an audio device or telephone.
- Interference from a nearby CW transmitter can produce on-and-off humming or clicking in an audio device or telephone system.
- Assuming the equipment is connected to a ground rod, if you receive an RF burn when touching your equipment while transmitting on an HF band, the ground wire might have high impedance on that frequency.
- A resonant ground connection can cause high RF voltages on the enclosures of station equipment.
- A good way to avoid unwanted effects of stray RF energy in an amateur station is to connect all equipment grounds together.
- Placing a ferrite bead around an audio cable will reduce RF interference caused by common-mode current.
- A ground loop can be avoided by connecting all ground conductors to a single point.
- Receiving reports of "hum" on your transmitted signal may indicate a ground loop somewhere in your station.
- A digital signal processor (DSP) can be used to remove noise from received signals in an amateur station.
- Compared to an analog filter, an advantage of a receiver DSP IF filter is the wide range of filter bandwidths and shapes that can be created.
- A DSP filter can perform automatic notching of interfering carriers.

Group G4D – Amateur Radio Practices

Questions in group G4D cover the following topics: speech processors, S meters, and sideband operation near band edges.

G4D Test Notes

- The speech processor in modern transceivers increases the intelligibility of transmitted phone signals during poor conditions.
- A speech processor increases average power in a transmitted single-sideband phone signal.

- An incorrectly adjusted speech processor can result in distorted speech, splatter, and excessive background pickup.
- An S meter measures received signal strength.
- Assuming a properly calibrated S meter, a reading of 20 dB over s-9 is 100 times stronger than an s-9 signal.
- An S meter is found in a receiver.
- The power output of a transmitter must be raised approximately four times to change the S-meter reading on a distant receiver from S8 to S9.
- When the displayed carrier frequency is set to 7.178 MHz, a 3 kHz LSB signal occupies the frequency range of 7.175 to 7.178 MHz.
- When the displayed carrier frequency is set to 14.347 MHz, a 3 kHz USB signal occupies the frequency range of 14.347 to 14.350 MHz.
- When using 3 kHz wide LSB, the lower edge of your displayed carrier frequency must be at least 3 kHz above the edge of the 40 meter General Class phone segment.
- When using 3 kHz wide USB, the upper edge of your displayed carrier frequency must be at least 3 kHz below the edge of the 20 meter General Class band.

Group G4E – Amateur Radio Practices

Questions in group G4E cover HF mobile radio installations, plus emergency and battery-powered operation.

G4E Test Notes

- A mobile antenna capacitance hat is a device to electrically lengthen a short antenna.
- A corona ball on an HF mobile antenna reduces high voltage discharge from the tip.
- A direct, fused power connection to the battery with heavy gauge wire is ideal for a 100-watt HF mobile installation.
- It is best not to draw the DC power for a 100-watt HF transceiver from an automobile's auxiliary power socket, because the socket's wiring may be inadequate for the current drawn by the transceiver.

- The effectiveness of an HF mobile transceiver operating in the 75 meter band is most limited by the antenna system.
- One disadvantage of using a shortened mobile antenna, as opposed to a full-size antenna, is the operating bandwidth may be very limited.
- The vehicle control computer is most likely to cause interfering signals to be heard in the receiver of an HF mobile installation in a recent-model vehicle.
- Photovoltaic conversion is the process by which sunlight is changed directly into electricity.
- The open-circuit voltage from a modern, well-illuminated photovoltaic cell is approximately 0.5V DC.
- A series diode is connected between a solar panel and storage battery because the diode prevents self-discharge of the battery though the panel during times of low or no sunlight.
- A disadvantage of using wind as the primary source of power for an emergency station is that a large energy storage system is needed to supply power when there is no wind.

4.5 SUBELEMENT G5

Subelement G5 contains three targeted question groups with a total of 42 potential test questions. Three of the 35 [Element 3] test questions will be drawn from subelement G5.

Group G5A – Electrical Principles

Questions in group G5A cover the following topics: reactance, inductance, capacitance, impedance, and impedance matching.

G5A Test Notes

- Impedance is the opposition to the flow of current in an AC circuit.
- Reactance is the opposition to the flow of alternating current caused by capacitance or inductance.

- Opposition to the flow of alternating current in an inductor is caused by reactance.
- Opposition to the flow of alternating current in a capacitor is caused by reactance.
- As the frequency of the applied AC increases, inductive reactance increases.
- As the frequency of the applied AC increases, capacitive reactance decreases.
- When the impedance of an electrical load is equal to the internal impedance of the power source, the source can deliver maximum power to the load.
- Impedance matching is important so the source can deliver maximum power to the load.
- Ohm is the unit used to measure reactance.
- Ohm is the unit used to measure impedance.
- One method of impedance matching between two AC circuits is to insert an LC network between the two circuits.
- One reason to use an impedance matching transformer is to maximize the transfer of power.
- A transformer, a pi network, and a length of transmission line can all be used for impedance matching at radio frequencies.

Group G5B – Electrical Principles

Questions in group G5B cover the following topics: the decibel, current and voltage dividers, electrical power calculations, sine wave root-mean-square (RMS) values, and PEP calculations.

G5B Test Notes

- A two-times increase or decrease in power results in a change by approximately 3 dB.
- Total current equals the sum of currents through each branch of a parallel circuit.

- If 400V DC is supplied to an 800-ohm load, 200 watts of electrical power are used.
- A 12V DC light bulb that draws 0.2 amperes uses 2.4 watts of electrical power.
- When a current of 7.0 milliamperes flows through 1.25 kilohms, approximately 61 milliwatts are dissipated.
- The output PEP from a transmitter with 200 volts peak-to-peak across a 50-ohm dummy load equals 100 watts.
- The RMS value of an AC signal results in the same power dissipation as a DC voltage of the same value.
- The peak-to-peak voltage of a sine wave that has an RMS voltage of 120 volts equals 339.4 volts.
- The RMS voltage of a sine wave with a value of 17 volts peak equals 12 volts.
- A transmission line loss of 1 dB would result in a 20.5% power loss.
- The ratio of peak envelope power to average power for an unmodulated carrier is 1.00.
- The RMS voltage across a 50-ohm dummy load dissipating 1200 watts equals 245 volts.
- If an average reading wattmeter connected to a transmitter output indicates 1060 watts, the output PEP of the unmodulated carrier equals 1060 watts.
- If an oscilloscope measures 500 volts peak-to-peak across a 50-ohm resistor connected to a transmitter output, the output PEP equals 625 watts.

Group G5C – Electrical Principles

Questions in group G5C cover the following topics: resistors, capacitors, and inductors in series and parallel; plus transformers.

G5C Test Notes

- Mutual inductance causes a voltage to appear across the secondary winding of a transformer when an AC voltage source is connected across its primary winding.

- The primary of a transformer is normally connected to the incoming source of energy.
- A resistor should be added in series to an existing resistor to increase resistance.
- The total resistance of three 100-ohm resistors in parallel equals 33.3 ohms.
- Three 150-ohm resistors in parallel produce 50 ohms of resistance, and the same three resistors in series produce 450 ohms.
- If the 2250-turn primary in a transformer is connected to 120V AC, the RMS voltage across a 500-turn secondary winding equals 26.7 volts.
- The turns ratio of a transformer used to match an audio amplifier having a 600-ohm output impedance to a speaker having a 4-ohm impedance equals 12.2 to 1.
- The total equivalent capacitance of two 5000-picofarad capacitors and one 750-picofarad capacitor connected in parallel equals 10750 picofarads.
- The total equivalent capacitance of three 100-microfarad capacitors connected in series equals 33.3 microfarads.
- The total equivalent inductance of three 10-millihenry inductors connected in parallel equals 3.3 millihenrys.
- The inductance of a 20-millihenry inductor in series with a 50-millihenry inductor equals 70 millihenrys.
- The capacitance of a 20-microfarad capacitor in series with a 50-microfarad capacitor equals 14.3 microfarads.
- A capacitor should be added in parallel with another capacitor to increase capacitance.
- An inductor should be added in series with another inductor to increase inductance.
- The total equivalent resistance of a 10-ohm, 20-ohm, and 50-ohm resistor in parallel equals 5.9 ohms.

4.6 SUBELEMENT G6

Subelement G6 contains three targeted question groups with a total of 46 potential test questions. Three of the 35 [Element 3] test questions will be drawn from subelement G6.

Group G6A – Circuit Components

Questions in group G6A cover resistors, capacitors, and inductors.

G6A Test Notes

- Low equivalent series resistance is an important characteristic for capacitors used to filter the DC output of a switching power supply.
- Electrolytic capacitors are often used in power supply circuits to filter the rectified AC.
- Ceramic capacitors generally cost less than other types of capacitors.
- An advantage of an electrolytic capacitor is high capacitance for given volume.

- One effect of lead inductance in a capacitor used at VHF and above is the effective capacitance may be reduced.
- If the temperature of a resistor is increased, the resistance will change depending on the resistor's temperature coefficient.
- A reason not to use wire-wound resistors in an RF circuit is the resistor's inductance could make circuit performance unpredictable.
- A thermistor is a device having a specific change in resistance with temperature variations.
- Advantages of using a ferrite core toroidal inductor are large values of inductance may be obtained, magnetic properties of the core may be optimized for a specific range of frequencies, and most of the magnetic field is contained in the core.
- The winding axes of solenoid inductors should be placed at right angles to minimize their mutual inductance.
- Minimizing the mutual inductance between two inductors reduces unwanted coupling between circuits.
- Filter choke is a common name for an inductor used to help smooth the DC output from the rectifier in a conventional power supply.
- Inter-turn capacitance in an inductor can result in the inductor becoming self-resonant at some frequencies.

Group G6B – Circuit Components

Questions in group G6B cover the following topics: rectifiers, solid-state diodes and transistors, vacuum tubes, and batteries.

G6B Test Notes

- The peak-inverse-voltage rating of a rectifier is the maximum voltage the rectifier will handle in the nonconducting direction.
- Two major ratings that must not be exceeded for silicon diode rectifiers are peak inverse voltage and average forward current.
- The approximate junction threshold voltage of a germanium diode is 0.3 volts.

- When two or more diodes are connected in parallel to increase current handling capacity, the purpose of the resistor connected in series with each diode is to ensure that one diode doesn't carry most of the current.
- The approximate junction threshold voltage of a conventional silicon diode is 0.7 volts.
- An advantage of using a Schottky diode in an RF switching circuit as compared to a standard silicon diode is lower capacitance.
- The stable operating points for a bipolar transistor used as a switch in a logic circuit are its saturation and cut-off regions.
- The cases of some large power transistors must be insulated from ground to avoid shorting the collector or drain voltage to ground.
- The gate of a MOSFET is separated from the channel with a thin insulating layer.
- The control grid element of a triode vacuum tube is used to regulate the flow of electrons between cathode and plate.
- A field-effect transistor is like a vacuum tube in its general operating characteristics.
- The primary purpose of a screen grid in a vacuum tube is to reduce grid-to-plate capacitance.
- An advantage of low internal resistance in nickel-cadmium batteries is high discharge currents.
- The minimum allowable discharge voltage for maximum life of a standard 12-volt lead-acid battery is 10.5 volts.
- It is NEVER acceptable to recharge a carbon-zinc primary cell.

Group G6C – Circuit Components

Questions in group G6C cover the following topics: analog and digital integrated circuits (ICs), microprocessors, memory, I/O devices, microwave ICs (MMICs), plus display devices.

G6C Test Notes

- A linear voltage regulator is an analog integrated circuit.
- The term MMIC stands for monolithic microwave integrated circuit.

- An advantage of CMOS-integrated circuits compared to TTL-integrated circuits is low power consumption.
- The term ROM stands for read-only memory.
- When memory is characterized as "non-volatile," the stored information is maintained even if power is removed.
- Integrated circuit operational amplifiers are analog.
- One disadvantage of an incandescent indicator compared to an LED is high power consumption.
- An LED is forward biased when emitting light.
- A liquid crystal display requires ambient or backlighting.
- An Amateur Radio station computer and transceiver may be connected using a USB interface.
- A microprocessor is a computer on a single integrated circuit.
- DE-9 connectors are a good choice for serial data ports.
- PL-259 connectors are commonly used for RF service at frequencies up to 150 MHz.
- RCA Phono connectors are commonly used for audio signals in Amateur Radio stations.
- The main reason to use keyed connectors instead of non-keyed types is to reduce the chance of incorrect mating.
- A type-N connector is a moisture-resistant RF connector useful to 10 GHz.
- DIN-type connectors are a family of multiple circuit connectors suitable for audio and control signals.
- SMA connectors are small threaded connectors suitable for signals up to several GHz.

4.7 SUBELEMENT G7

Subelement G7 contains three targeted question groups with a total of 38 potential test questions. Three of the 35 [Element 3] test questions will be drawn from subelement G7.

Group G7A – Practical Circuits

Questions in group G7A cover power supplies and schematic symbols. Several questions from this group are based on the schematic diagram shown below.

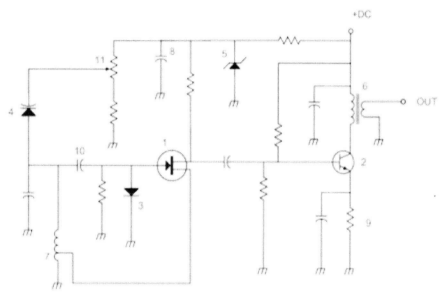

Figure G7-1

G7A Test Notes

- Power-supply bleeder resistors discharge filter capacitors for safety.
- Capacitors and inductors are used for power-supply filter networks.
- The peak inverse voltage across the rectifiers in a full-wave bridge power supply is equal to the normal peak output voltage of the power supply.
- The peak inverse voltage across the rectifier in a half-wave power supply is two times the normal peak output voltage of the power supply.
- 180 degrees of an AC cycle is converted to DC by a half-wave rectifier.
- 360 degrees of an AC cycle is converted to DC by a full-wave rectifier.

- The output waveform of an unfiltered full-wave rectifier connected to a resistive load is a series of DC pulses at twice the frequency of the AC input.
- An advantage of a switch-mode power supply as compared to a linear power supply is that high frequency operation allows the use of smaller components.
- Symbol 1 in Figure G7-1 represents a field-effect transistor.
- Symbol 5 in Figure G7-1 represents a Zener diode.
- Symbol 2 in Figure G7-1 represents an NPN junction transistor.
- Symbol 6 in Figure G7-1 represents a multiple-winding transformer.
- Symbol 7 in Figure G7-1 represents a tapped inductor.

Group G7B – Practical Circuits

Questions in group G7B cover digital circuits, amplifiers, and oscillators.

G7B Test Notes

- Complex digital circuitry can often be replaced by a microcontroller integrated circuit.
- An advantage of using the binary system when processing digital signals is binary "ones" and "zeros" are easy to represent with an "on" or "off" state.
- The output of a two input AND gate is high only when both inputs are high.
- The output of a two input NOR gate is low when either or both inputs are high.
- A 3-bit binary counter has eight states.
- A shift register is a clocked array of circuits that passes data in steps along the array.
- The basic components of virtually all sine wave oscillators are a filter and an amplifier operating in a feedback loop.
- The efficiency of an RF power amplifier is determined by dividing the RF output power by the DC input power.

- The inductance and capacitance in the tank circuit determine the frequency of an LC oscillator.
- A characteristic of a Class A amplifier is its low distortion.
- CW mode is a Class C power stage appropriate for amplifying a modulated signal.
- Class C amplifiers have the highest efficiency.
- Neutralizing the final amplifier stage of a transmitter eliminates self-oscillations.
- A linear amplifier is an amplifier in which the output preserves the input waveform.

Group G7C – Practical Circuits

Questions in group G7C cover receivers, transmitters, filters, and oscillators.

G7C Test Notes

- A filter is used to process signals from the balanced modulator and send them to the mixer in a single-sideband phone transmitter.
- A balanced modulator circuit is used to combine signals from the carrier oscillator and speech amplifier, and send the result to the filter in a typical single-sideband phone transmitter.
- A mixer circuit is used to process signals from the RF amplifier and local oscillator, and send the result to the IF filter in a superheterodyne receiver.
- A product detector circuit is used to combine signals from the IF amplifier and BFO, and send the result to the AF amplifier in a single-sideband receiver.
- A transceiver controlled by a direct digital synthesizer (DDS) has the advantage of variable frequency with the stability of a crystal oscillator.
- The impedance of a low-pass filter and the impedance of the transmission line into which it is inserted should be about the same.
- The simplest combination of stages that implement a superheterodyne receiver is HF oscillator, mixer, and detector.
- A discriminator circuit is used in many FM receivers to convert signals coming from the IF amplifier to audio.

- An analog to digital converter, a digital to analog converter, and a digital processor chip are all needed for a digital signal processor IF filter.
- Digital signal processor filtering is accomplished by converting the signal from analog to digital and using digital processing.
- The term "software-defined radio" (SDR) refers to a radio in which most major signal processing functions are performed by software.

4.8 SUBELEMENT G8

Subelement G8 contains two targeted question groups with a total of 24 potential test questions. Two of the 35 [Element 3] test questions will be drawn from subelement G8.

Group G8A – Carriers and Modulation

Questions in group G8A cover the following topics: AM, FM, single and double sideband, modulation envelope, and overmodulation.

G8A Test Notes

- Amplitude modulation is the process that changes the envelope of an RF wave to convey information.
- Phase modulation is the process that changes the phase angle of an RF wave to convey information.
- Frequency modulation is the process that changes the frequency of an RF wave to convey information.
- A reactance modulator connected to an RF power amplifier produces phase modulation.
- Amplitude modulation varies the instantaneous power level of the RF signal.
- An advantage of carrier suppression in a single-sideband phone transmission is the available transmitter power can be used more effectively.
- Single-sideband phone emissions use the narrowest frequency bandwidth.

- Excessive bandwidth is an effect of overmodulation.
- The transmit audio or microphone gain control is typically adjusted for proper ALC setting on an amateur single-sideband transceiver.
- Flat-topping of a single-sideband phone transmission is the signal distortion caused by excessive drive.
- When a modulating audio signal is applied to an FM transmitter, the RF carrier signal changes proportionally to the instantaneous amplitude of the modulating signal.
- Both upper and lower sideband signals are found at the output of a properly adjusted balanced modulator.

Group G8B – Signals and Emission

Questions in group G8B cover the following topics: frequency mixing, multiplication, HF data communications, bandwidths of various modes, and deviation.

G8B Test Notes

- A receiver's mixer stage can combine a 14.250 MHz input signal with a 13.795 MHz oscillator signal to produce a 455 kHz intermediate frequency (IF) signal.
- If a receiver mixes a 13.800 MHz VFO with a 14.255 MHz received signal to produce a 455 kHz IF signal, a 13.345 MHz signal will produce image response interference in the receiver.
- "Heterodyning" is another term for the mixing of two RF signals.
- The multiplier stage in a VHF FM transmitter generates a harmonic of a lower frequency signal to reach the desired operating frequency.
- Frequency modulated (FM) phone is not used below 29.5 MHz because the wide bandwidth is prohibited by FCC rules.
- The total bandwidth of an FM-phone transmission having a 5 kHz deviation and a 3 kHz modulating frequency is 16 kHz.
- The frequency deviation for a 12.21-MHz reactance-modulated oscillator in a 5-kHz deviation, 146.52-MHz FM-phone transmitter is 416.7 Hz.

- It is important to know the duty cycle of the data mode you are using when transmitting, because some modes have high duty cycles, which could exceed the transmitter's average power rating.
- Matching receiver bandwidth to the bandwidth of the operating mode results in the best signal to noise ratio.
- The number 31 in PSK31 represents the approximate transmitted symbol rate.
- Forward error correction allows the receiver to correct errors in received data packets by transmitting redundant information with the data.
- Higher transmitted symbol rates require higher bandwidth.

4.9 SUBELEMENT G9

Subelement G9 contains four targeted question groups with a total of 56 potential test questions. Four of the 35 [Element 3] test questions will be drawn from subelement G9.

Group G9A – Antennas and Feed Lines

Questions in group G9A cover the following topics: feed line characteristic impedance and attenuation, SWR calculation, measurement and effects, plus matching networks.

G9A Test Notes

- The distance between the center of the conductors and the radius of the conductors determine the characteristic impedance of a parallel conductor antenna feed line.
- The typical characteristic impedances of coaxial cables used for antenna feed lines at amateur stations are 50 and 75 ohms.
- The characteristic impedance of flat ribbon TV type twinlead is 300 ohms.
- Reflected power at the point where a feed line connects to an antenna is caused by the difference between feed-line impedance and antenna feed-point impedance.

- The attenuation of coaxial cable increases as the signal frequency it is carrying increases.
- RF feed line losses are usually expressed in dB per 100 ft.
- The antenna feed-point impedance must be matched to the characteristic impedance of the feed line to prevent standing waves.
- If the SWR on an antenna feed line is 5 to 1 and a matching network at the transmitter end of the feed line is adjusted to 1 to 1, the resulting feed line SWR is 5 to 1.
- Connecting a 50-ohm feed line to a non-reactive load with a 200-ohm impedance will result in a 4:1 standing-wave ratio.
- Connecting a 50-ohm feed line to a non-reactive load with a 10-ohm impedance will result in a 5:1 standing-wave ratio.
- Connecting a 50-ohm feed line to a non-reactive load having a 50-ohm impedance will result in a 1:1 standing-wave ratio.
- Feeding a vertical antenna that has a 25-ohm feed-point impedance with 50-ohm coaxial cable will result in a 2:1 SWR.
- Feeding an antenna that has a 300-ohm feed-point impedance with 50-ohm coaxial cable will result in a 6:1 SWR.

Group G9B – Antennas and Feed Lines

Questions in group G9B cover basic antennas.

G9B Test Notes

- A disadvantage of directly fed random-wire antennas is that you may experience RF burns when touching metal objects in your station.
- An advantage of downward sloping radials on a 1/4-wave ground-plane antenna is that they bring the feed-point impedance closer to 50 ohms.
- When its radials are changed from horizontal to downward sloping, the feed-point impedance of a ground-plane antenna increases.
- The low angle azimuthal radiation pattern of an ideal 1/2-wavelength dipole antenna, installed 1/2 wavelength high and parallel to the Earth, is a figure eight at right angles to the antenna.

- If a horizontal dipole HF antenna is less than 1/2 wavelength high, the horizontal (azimuthal) radiation pattern is almost omnidirectional.
- The radial wires of a ground-mounted vertical antenna system should be placed on the surface or buried a few inches below the ground.
- The feed-point impedance of a 1/2-wavelength dipole antenna steadily decreases as the antenna is lowered from 1/4 wave above ground.
- The feed-point impedance of a 1/2-wavelength dipole steadily increases as the feed-point location is moved from the center toward the ends.
- An advantage of a horizontally polarized HF antenna, as compared to vertically polarized, is lower ground reflection losses.
- The length for a 1/2-wave dipole antenna cut for 14.250 MHz is approximately 32 feet.
- The length for a 1/2-wave dipole antenna cut for 3.550 MHz is approximately 131 feet.
- The length for a 1/4-wave vertical antenna cut for 28.5 MHz is approximately 8 feet.

Group G9C – Antennas and Feed Lines

Questions in group G9C cover directional antennas.

G9C Test Notes

- Larger diameter elements will increase the bandwidth of a Yagi antenna.
- The approximate length of the driven element of a Yagi antenna is 1/2 wavelength.
- The director of a three-element, single-band Yagi antenna is normally the shortest parasitic element
- The reflector of a three-element, single-band Yagi antenna is normally the longest parasitic element.
- Increasing boom length and adding directors to a Yagi antenna will increase its gain.
- A Yagi antenna is often used for radio communications on the 20 meter band because it helps reduce interference from other stations to the side or behind the antenna.

- "Front-to-back ratio" of a Yagi antenna is the power radiated in the major radiation lobe compared to the power radiated in exactly the opposite direction.
- The "main lobe" of a directive antenna is the direction of maximum radiated field strength from the antenna.
- The maximum theoretical forward gain of a three-element, single-band Yagi antenna is approximately 9.7 dBi.
- Physical length of the boom, number of elements on the boom, and spacing of each element along the boom are all Yagi antenna design variables that can be adjusted to optimize forward gain, front-to-back ratio, or SWR bandwidth.
- The purpose of a gamma match used with Yagi antennas is to match the relatively low feed-point impedance to 50 ohms.
- An advantage of using a gamma match for impedance matching of Yagi antennas to 50-ohm coax feed line is that it does not require the elements be insulated from the boom.
- Each side of a quad antenna driven element is approximately 1/4 wavelength.
- The forward gain of a two-element quad antenna is about the same as the forward gain of a three-element Yagi antenna.
- Each side of a quad antenna reflector element is slightly longer than 1/4 wavelength.
- The gain of a two-element delta-loop beam is about the same as the gain of a two-element quad antenna.
- Each leg of a symmetrical delta-loop antenna is approximately 1/3 wavelength.
- When the feed point of a quad antenna is changed from the center of either horizontal wire to the center of either vertical wire, the polarization of the radiated signal changes from horizontal to vertical.
- To operate as a beam antenna, the reflector element of a two-element quad antenna must be approximately 5% longer than the driven element.

- The gain of two three-element horizontally polarized Yagi antennas (spaced vertically 1/2 wavelength apart) is approximately 3 dB higher than the gain of a single three-element Yagi.

Group G9D – Antennas and Feed Lines

Questions in group G9D cover specialized antennas.

G9D Test Notes

- The term "NVIS" stands for near vertical incidence sky-wave.
- An advantage of an NVIS antenna is high vertical angle radiation for working stations within a radius of a few hundred kilometers.
- An NVIS antenna is typically installed between 1/10 and 1/4 wavelength above ground.
- The primary purpose of antenna traps is to permit multiband operation.
- The advantage of vertically stacking horizontally polarized Yagi antennas is that it narrows the main lobe in elevation.
- An advantage of a log periodic antenna is wide bandwidth.
- The length and spacing of the elements on a log periodic antenna increase logarithmically from one end of the boom to the other.
- A Beverage antenna is not used for transmitting because it has high losses compared to other types of antennas.
- Beverage antennas are often used for directional receiving in low HF bands.
- A Beverage antenna is a very long and low directional receiving antenna.
- Multiband antennas have poor harmonic rejection.

4.10 SUBELEMENT G0

Subelement G0 contains two targeted question groups with a total of 28 potential test questions. Two of the 35 [Element 3] test questions will be drawn from subelement G0.

Group G0A – Electrical and RF Safety

Questions in group G0A cover RF safety principles, rules and guidelines, and routine station evaluation.

G0A Test Notes

- RF energy affects the human body by heating body tissue.
- Duty cycle, frequency, and power density are all important in estimating whether an RF signal exceeds the maximum permissible exposure (MPE).
- Calculation based on FCC OET Bulletin 65, calculation based on computer modeling, and measurement of field strength using calibrated equipment are all methods of determining whether a station complies with FCC RF exposure regulations.
- In reference to RF radiation exposure, the term "time averaging" is the total RF exposure averaged over a certain time.
- If the RF energy radiated from your station exceeds permissible limits, you must take action to prevent human exposure to the excessive RF fields.
- A lower transmitter duty cycle permits greater short-term exposure levels.
- When transmitter power exceeds levels specified in Part 97.13, amateur operators must perform a routine RF exposure evaluation to ensure compliance with safety regulations.
- A calibrated field-strength meter with a calibrated antenna can be used to accurately measure an RF field.
- If a neighbor might receive more than the allowable limit of RF exposure from the main lobe of a directional antenna, take precautions to ensure that the antenna cannot be pointed in their direction.

- If installing an indoor transmitting antenna, you should make sure that MPE limits are not exceeded in occupied areas.
- Turn off the transmitter and disconnect the feed line whenever you make adjustments or repairs to an antenna.
- A ground-mounted antenna should be installed so no one can be exposed to RF radiation in excess of maximum permissible limits.

Group G0B – Electrical and RF Safety

Questions in group G0B cover the following topics: electrical shock and treatment, safety grounding, fusing, interlocks, wiring, plus antenna and tower safety.

G0B Test Notes

- For devices operated from a 240V AC single-phase source, only the hot wires in a four-conductor line cord should be attached to fuses or circuit breakers.
- AWG number 12 is the minimum wire size that may be safely used for a circuit that draws up to 20 amperes of continuous current.
- Use 15-ampere fuses or circuit breakers for circuits on AWG number 14 wiring.
- Due to the danger of carbon monoxide poisoning, DO NOT place a gasoline-fueled generator inside an occupied area.
- Current flowing from one or more of the hot wires directly to ground will cause a ground fault circuit interrupter (GFCI) to disconnect the AC line power from a device.
- The metal enclosure of ALL station equipment must be grounded to ensure hazardous voltages cannot appear on the chassis.
- When climbing on a tower using a safety belt or harness, always attach the belt safety hook to the belt D ring with the hook opening away from the tower.
- When preparing to climb a tower that supports electrically powered devices, make sure all circuits that supply power to the tower are locked out and tagged.

- Soldered joints should not be used with the wires that connect the base of a tower to a ground rod system because they will likely be destroyed by the heat of a lightning strike.
- The lead found in lead-tin solder can contaminate food if hands are not washed carefully after handling.
- Lightning protection grounds must be bonded together with all other grounds.
- A transmitter power supply interlock ensures that dangerous voltages are removed if the cabinet is opened.
- Disconnect the incoming utility power feed when powering your house from an emergency generator.
- The National Electrical Code covers electrical safety inside the ham shack.
- An emergency generator should be located in a well-ventilated area.
- Lead-acid storage batteries can give off explosive hydrogen gas when being charged.

Study Tips

Review the test notes for each subelement carefully. Once you are comfortable with the majority of information presented, take the practice exam in Appendix 2 to check your progress. Repeat the process as needed.

You should also try a random sampling of the Element 3 question pool for more practice. The resource directory in Appendix 4 will show you not only where to download the entire question pool online but also where to find detailed explanations for all material covered on the licensing exam. Everything you need to pass your Amateur Radio (Ham) License Examination is in this manual – all you have to do is use the information, study with diligence, and practice!

5. AMATEUR EXTRA CLASS (ELEMENT 4)

Amateur Extra Class written examinations are referred to as Element 4. Topics covered during examination include: FCC rules, operating procedures, radio wave propagation, amateur practices, electrical principles, circuit components, practical circuits, signals and emissions, antennas and transmission lines, and safety.

Element 4 exams are derived from 10 subelements containing 50 targeted question groups. The current public question pool contains a total of 702 available test questions. Element 4 tests are made up of 50 random questions, of which at least 37 must be answered correctly to qualify for an Amateur Extra Class operator license. Following is a rundown of applicable subelements and related question groups.

5.1 SUBELEMENT E1

Subelement E1 contains six targeted question groups with a total of 74 potential test questions. Six of the 50 [Element 4] test questions will be drawn from subelement E1.

Group E1A – Operating Standards

Questions in group E1A cover the following topics: frequency privileges, emission standards, automatic message forwarding, frequency sharing, and stations aboard ships or aircraft.

E1A Test Notes

- When using a transceiver that displays the carrier frequency of phone signals, 3 kHz below the upper band edge is the highest frequency at which a properly adjusted USB emission will be totally within the band.
- When using a transceiver that displays the carrier frequency of phone signals, 3 kHz above the lower band edge is the lowest frequency at which a properly adjusted LSB emission will be totally within the band.
- With your transceiver displaying the carrier frequency of phone signals, it is NOT legal to use the same frequency to return a DX station's CQ on 14.349 MHz USB, because your sidebands will extend beyond the band edge.
- With your transceiver displaying the carrier frequency of phone signals, it is NOT legal to use the same frequency to return a DX station's CQ on 3.601 MHz LSB, because your sidebands will extend beyond the edge of the phone band segment.
- 100 watts PEP effective radiated power relative to an isotropic radiator is the maximum power output permitted on the 60 meter band.
- Operation is restricted to specific emission types and channels on the 60 meter band.
- 60 meter is the only amateur band where transmission on specific channels rather than a range of frequencies is permitted.

- If a station in a message forwarding system inadvertently forwards a message that is in violation of FCC rules, the control operator of the originating station is primarily accountable for the rules violation.
- If your digital message forwarding station inadvertently forwards a communication that violates FCC rules, immediately discontinue forwarding the communication as soon as you become aware of it.
- An amateur station aboard a ship or aircraft must be approved by the master of the ship or the pilot in command of the aircraft before the station is operated.
- Any FCC-issued amateur license or a reciprocal permit for an alien amateur license is required when operating an amateur station aboard a U.S.-registered vessel in international waters.
- With your transceiver displaying the carrier frequency of CW signals, it is NOT legal to use the same frequency to return a DX station's CQ on 3.500 MHz, because the sidebands from the CW signal will be out of the band.
- Any person holding an FCC-issued amateur license or who is authorized for alien reciprocal operation must be in physical control of the station apparatus of an amateur station aboard any vessel or craft that is documented or registered in the United States.

Group E1B – Restrictions and Special Operations

Questions in group E1B cover the following topics: restrictions on station location, general operating restrictions, spurious emissions, control operator reimbursement, antenna structure restrictions, plus RACES operations.

E1B Test Notes

- A spurious emission is an emission outside its necessary bandwidth that can be reduced or eliminated without affecting the information transmitted.
- The physical location of an amateur station apparatus or antenna structure could be restricted if it is of environmental importance or significant in American history, architecture, or culture.

- Amateur stations must protect FCC monitoring facilities within 1 mile of the station from harmful interference.
- An Environmental assessment must be submitted to the FCC before placing an amateur station within an officially designated wilderness area or preserve, or within an area listed in the National Register of Historical Places.
- 2.8 kHz is the maximum bandwidth for a data emission on 60 meters.
- If you are installing an amateur station antenna at a site at or near a public use airport, you may have to notify the Federal Aviation Administration and register it with the FCC as required by Part 17 of FCC rules.
- The carrier frequency of a CW signal must be set at the center frequency of the channel to comply with FCC rules for 60 meter operation.
- If an amateur station signal causes interference to domestic broadcast reception, the FCC may limit the station to avoid transmitting during certain hours on frequencies that cause the interference.
- Any FCC-licensed amateur station certified by the responsible civil defense organization for the area served may be operated in RACES.
- All amateur service frequencies authorized to the control operator are authorized to an amateur station participating in RACES.
- The permitted mean power of any spurious emission is at least 43 dB below the mean power of the fundamental emission from a station transmitter or external RF amplifier installed after January 1, 2003, and transmitting on a frequency below 30 MHz.
- 1.0 is the highest modulation index permitted at the highest modulation frequency for angle modulation.

Group E1C – Station Control

Questions in group E1C cover the following topics: definitions and restrictions, automatic and remote control operation, and control operator responsibilities.

E1C Test Notes

- A remotely controlled station is a station controlled indirectly through a control link.
- Automatic control of a station is the use of devices and procedures for control so that the control operator does not have to be present at a control point.
- Unlike a station under local control, the control operator of a station under automatic control is not required to be present at the control point.
- An automatically controlled station may retransmit third-party communications only when transmitting RTTY or data emissions.
- An automatically controlled station may NEVER originate third-party communications.
- A control operator must be present at the control point of remotely controlled amateur stations.
- Local control is direct manipulation of the transmitter by a control operator.
- Three minutes is the maximum permissible duration of a remotely controlled station's transmissions if its control link malfunctions.
- 29.500 – 29.700 MHz are the available frequencies for automatically controlled repeaters operating below 30 MHz.
- Only auxiliary, repeater, or space stations may automatically retransmit the radio signals of other amateur stations.

Group E1D – Amateur Satellite Service

Questions in group E1D cover the following topics: definitions and purpose, license requirements for space stations, available frequencies and bands, telecommand and telemetry operations, restrictions, and special provisions, and notification requirements.

E1D Test Notes

- Telemetry is one-way transmission of measurements at a distance from the measuring instrument.

- The amateur satellite service is a radio communications service using amateur radio stations on satellites.
- A telecommand station is an amateur station that transmits communications to initiate, modify, or terminate functions of a space station.
- An Earth station is an amateur station within 50 km of the Earth's surface intended for communications with amateur stations by means of objects in space.
- All classes of licensee are authorized to be the control operator of a space station.
- A space station must be capable of terminating transmissions by telecommand when directed by the FCC.
- Only 40, 20, 17, 15, 12 and 10 meter amateur service HF bands have frequencies authorized to space stations.
- 2 meter is the only amateur service VHF band with frequencies available for space stations.
- Only 70 cm, 23 cm, and 13 cm amateur service UHF bands have frequencies available for space stations.
- Any amateur station so designated by the space station licensee, subject to the privileges of the class of operator license held by the control operator, is eligible to be a telecommand station.
- Any amateur station, subject to the privileges of the class of operator license held by the control operator, is eligible to operate as an Earth station.

Group E1E – Volunteer Examiner Program

Questions in group E1E cover the following topics: definitions, qualifications, preparation and administration of exams, accreditation, question pools, and documentation requirements.

E1E Test Notes

- A minimum of three qualified VEs are required to administer an Element 4 amateur operator license examination.
- The questions for all written U.S. amateur license examinations are listed in a question pool maintained by all the VECs.
- A Volunteer Examiner Coordinator (VEC) is an organization that has entered into an agreement with the FCC to coordinate amateur operator license examinations.
- The Volunteer Examiner (VE) accreditation process is the procedure by which a VEC confirms that the VE applicant meets FCC requirements to serve as an examiner.
- The minimum passing score on amateur operator license examinations is 74%.
- Each administering VE is responsible for the proper conduct and necessary supervision during an amateur operator license examination session.
- If a candidate fails to comply with the examiner's instructions during an amateur operator license examination, a VE should immediately terminate the candidate's examination.
- As listed in the FCC rules, VEs may not administer an examination to their relatives.
- Penalty for a VE who fraudulently administers or certifies an examination is revocation of the VE's amateur station license grant and the suspension of the VE's amateur operator license grant.
- After administration of a successful examination for an amateur operator license, the administering VEs must submit the application document according to the coordinating VEC instructions.
- If an examinee scores a passing grade on all examination elements needed for an upgrade or new license, three VEs must certify that the examinee is qualified for the license grant and that they have complied with the administering VE requirements.

- If an examinee does not pass the exam, the VE team must return the application document to the examinee.
- The licensee's license will be cancelled if he or she fails to appear for re-administration of an examination when so directed by the FCC.
- Part 97 rules state that VEs and VECs may be reimbursed for preparing, processing, administering and coordinating an examination for an amateur radio license.

Group E1F – Miscellaneous Rules

Questions in group E1F cover the following topics: external RF power amplifiers, national quiet zone, business communications, compensated communications, spread spectrum, auxiliary stations, reciprocal operating privileges, IARP and CEPT licenses, third-party communications with foreign countries, plus special temporary authority.

E1F Test Notes

- Spread spectrum transmissions are permitted only on amateur frequencies above 222 MHz.
- The CEPT agreement allows an FCC-licensed U.S. citizen to operate in many European countries, and alien amateurs from many European countries to operate in the United States.
- A dealer may sell an external RF power amplifier capable of operation below 144 MHz that has not been granted FCC certification if it was purchased in used condition from an amateur operator and is sold to another amateur operator for use at that operator's station.
- "Line A" is roughly parallel to and south of the U.S.-Canadian border.
- Amateur stations may not transmit in the 420 – 430 MHz frequency segment if they are located in the contiguous 48 states and north of Line A.
- The National Radio Quiet Zone is an area surrounding the National Radio Astronomy Observatory.

- An amateur station may send a message to a business when neither the amateur nor their employer has a pecuniary interest in the communications.
- Communications transmitted for hire or material compensation are prohibited, except as otherwise provided in the rules.
- When transmitting spread spectrum emission, a station must not cause harmful interference to other stations employing other authorized emissions, the station must be in an area regulated by the FCC or in a country that permits SS emissions, and the transmission must not be used to obscure the meaning of any communication.
- The maximum transmitter power for an amateur station transmitting spread spectrum communications is 10 watts.
- An external RF power amplifier must satisfy the FCC's spurious emission standards when operated at the lesser of 1500 watts, or its full output power.
- Only Technician, General, Advanced or Amateur Extra Class operators may be the control operator of an auxiliary station.
- Communications incidental to the purpose of the amateur service and remarks of a personal nature may be transmitted to amateur stations in foreign countries.
- The FCC might issue a "Special Temporary Authority" (STA) to an amateur station to provide for experimental amateur communications.

5.2 SUBELEMENT E2

Subelement E2 contains five targeted question groups with a total of 68 potential test questions. Five of the 50 [Element 4] test questions will be drawn from subelement E2.

Group E2A – Amateur Radio in Space

Questions in group E2A cover the following topics: amateur satellites, orbital mechanics, frequencies and modes, satellite hardware, and satellite operations.

E2A Test Notes

- An ascending pass for an amateur satellite is from south to north.
- A descending pass for an amateur satellite is from north to south.
- The orbital period of an Earth satellite is the time it takes for a satellite to complete one revolution around the Earth.
- The term "mode" refers to the type of signals that can be relayed through a satellite.
- The letters in a satellite's mode designator specify the uplink and downlink frequency ranges.
- If operating in mode U/V, a satellite would receive signals on the 435 – 438 MHz band.
- FM, CW, SSB, SSTV, PSK, and Packet signals can be relayed through a linear transponder.
- Effective radiated power to a satellite that uses a linear transponder should be limited to avoid reducing the downlink power to all other users.
- The terms "L band" and "S band" specify the 23 centimeter and 13 centimeter bands.
- The received signal from an amateur satellite may exhibit a rapidly repeating fading effect because the satellite is spinning.
- A circularly polarized antenna can be used to minimize the effects of spin modulation and Faraday rotation.

- One way to predict the location of a satellite at a given time is by calculations using the Keplerian elements for the specified satellite.
- A Geostationary satellite appears to stay in one position in the sky.

Group E2B – Television Practices

Questions in group E2B cover fast-scan television standards and techniques, plus slow-scan television standards and techniques.

E2B Test Notes

- In fast-scan (NTSC) television systems, a new frame is transmitted 30 times per second.
- A fast-scan (NTSC) television frame is made up of 525 horizontal lines.
- An interlaced scanning pattern in a fast-scan (NTSC) television system is generated by scanning odd-numbered lines in one field and even-numbered ones in the next.
- Blanking in a video signal is turning off the scanning beam while it is traveling from right to left or from bottom to top.
- Using vestigial sideband for standard fast-scan TV transmissions reduces bandwidth while allowing for simple video detector circuitry.
- Vestigial sideband modulation is amplitude modulation in which one complete sideband and a portion of the other are transmitted.
- The chroma signal component carries color information in NTSC video.
- Frequency-modulated sub-carrier, separate VHF or UHF audio link, and frequency modulation of the video carrier are all common methods of transmitting accompanying audio with amateur fast-scan television.
- Only a receiver with SSB capability and a suitable computer are needed to decode SSTV using Digital Radio Mondiale (DRM).
- 3 kHz is an acceptable bandwidth for Digital Radio Mondiale (DRM) based voice or SSTV digital transmissions made on the HF amateur bands.
- The vertical interval signaling (VIS) code transmitted as part of an SSTV transmission is to identify the SSTV mode being used.
- For analog SSTV images, varying tone frequencies representing the video are transmitted using single-sideband on the HF bands.

- 128 or 256 lines are commonly used in each frame on an amateur slow-scan color television picture.
- The tone frequency of an amateur slow-scan television signal encodes the brightness of the picture.
- Specific tone frequencies signal SSTV receiving equipment to begin a new picture line.
- NTSC is the video standard used by North American fast-scan amateur television (ATV) stations.
- The bandwidth of a slow-scan TV signal is approximately 3 kHz.
- One is likely to find FM ATV transmissions on 1255 MHz.
- Slow-scan TV transmissions are restricted to phone band segments, and their bandwidth can be no greater than that of a voice signal of the same modulation type.

Group E2C – Operating Methods

Questions in group E2C cover the following topics: contest and DX operating, spread spectrum transmissions, and selecting an operating frequency.

E2C Test Notes

- Contest operators are permitted to make contacts even if they do not submit a log.
- The term "self-spotting" refers to a prohibited practice of posting one's own call sign and frequency on a call sign spotting network during contest operation.
- Amateur radio contesting is generally excluded on the 30 meter band.
- Amateur radio contest contact is generally discouraged on 146.52 MHz.
- The function of a DX QSL Manager is to handle the receiving and sending of confirmation cards for a DX station.
- During a VHF/UHF contest, you can expect to find the highest level of activity in the weak signal segment of the band, with most of the activity near the calling frequency.
- The Cabrillo format is a standard for submission of electronic contest logs.

- Received spread spectrum signals are resistant to interference because the signals not using the spectrum-spreading algorithm are suppressed in the receiver.
- Frequency hopping is a spread spectrum technique in which a transmitted signal is changed very rapidly according to a particular sequence also used by the receiving station.
- A DX station might state that they are listening on another frequency to separate the calling stations from the DX station, to reduce interference (thereby improving operating efficiency), or because the DX station may be transmitting on a frequency that is prohibited to some responding stations.
- Send your full call sign once or twice when attempting to contact a DX station while working a pileup or in a contest.
- Switching to a lower frequency HF band might help restore contact when DX signals become too weak to copy across an entire HF band a few hours after sunset.

Group E2D – Operating Methods

Questions in group E2D cover VHF and UHF digital modes plus APRS.

E2D Test Notes
- FSK441 is especially designed for use for meteor scatter signals.
- Baud is defined as the number of data symbols transmitted per second.
- JT65 is especially useful for Earth-Moon-Earth (EME) communications.
- The digital store-and-forward functions on an amateur radio satellite are for storing digital messages in the satellite for later download by other stations.
- Low Earth-orbiting digital satellites normally use the store-and-forward technique to relay messages around the world.
- 144.39 MHz is a commonly used 2 meter APRS frequency.
- APRS uses AX.25 digital protocol.
- Unnumbered Information packet frames are used to transmit APRS beacon data.

- 300-baud packet has the fastest data throughput under clear communications conditions.
- An APRS station with a GPS unit can automatically transmit information to show a mobile station's position during a public service communications activity.
- Latitude and longitude data is used by the APRS network to communicate your location.
- JT65 improves EME communications by decoding signals many dB below the noise floor using forward error correction (FEC).

Group E2E – Operating Methods

Questions in group E2E cover the following topics: operating HF digital modes and error correction.

E2E Test Notes

- FSK modulation is common for data emissions below 30 MHz.
- FEC stands for forward error correction.
- Forward error correction is implemented by transmitting extra data that may be used to detect and correct transmission errors.
- Selective fading is when one of the ellipses in an FSK crossed-ellipse display suddenly disappears.
- ARQ accomplishes error correction by requesting a retransmission when errors are detected.
- 300 baud is the most common data rate used for HF packet communications.
- The typical bandwidth of a properly modulated MFSK16 signal is 316 Hz.
- PACTOR is an HF digital mode that can be used to transfer binary files.
- PSK31 is an HF digital mode that uses variable-length coding for bandwidth efficiency.
- PSK31 has the narrowest bandwidth of most digital modes.
- The difference between direct FSK and audio FSK is that direct FSK applies the data signal to the transmitter VFO.

- Winlink does not support keyboard-to-keyboard operation.

5.3 SUBELEMENT E3

Subelement E3 contains three targeted question groups with a total of 35 potential test questions. Three of the 50 [Element 4] test questions will be drawn from subelement E3.

Group E3A – Radio Wave Propagation

Questions in group E3A cover Earth-Moon-Earth communications and meteor scatter.

E3A Test Notes

- As long as both can "see" the Moon, the maximum separation measured along the surface of the Earth between two stations communicating by Moon bounce is approximately 12,000 miles.
- Libration fading of an Earth-Moon-Earth signal is a fluttery irregular fading.
- The least path loss for EME contacts will generally result when the moon is at perigee.
- Receiving equipment with very low noise figures is desirable for EME communications.
- A method of establishing EME contacts is time synchronous transmissions with each station alternating.
- EME signals in the 2 meter band are normally found at 144.000 – 144.100 MHz.
- EME signals in the 70 cm band are normally found at 432.000 – 432.100 MHz.
- When a meteor strikes the Earth's atmosphere, a cylindrical region of free electrons is formed at the E layer of the ionosphere.
- The 28 – 148 MHz frequency range is well suited for meteor-scatter communications.

- Fifteen-second timed transmission sequences with stations alternating based on location, use of high speed CW or digital modes, and short transmission with rapidly repeated call signs and signal reports are all good techniques for making meteor-scatter contacts.

Group E3B – Radio Wave Propagation

Questions in group E3B cover the following topics: transequatorial, long-path, gray-line, and multi-path propagation.

E3B Test Notes

- Transequatorial propagation is propagation between two midlatitude points at approximately the same distance north and south of the magnetic equator.
- The maximum range for transequatorial propagation is approximately 5000 miles.
- The best time of day for transequatorial propagation is afternoon or early evening.
- Long-path propagation requires an HF beam antenna be pointed in a direction 180 degrees away from a station to receive the strongest signals.
- 160 to 10 meter amateur bands support long-path propagation.
- The 20 meter amateur band most frequently provides long-path propagation.
- Receipt of a signal by more than one path could account for hearing an echo on the received signal of a distant station.
- Radio signals traveling along the terminator between daylight and darkness are normally associated with gray-line HF propagation.
- Gray-line propagation is most likely to occur at sunrise and sunset.
- Gray-line propagation occurs because D-layer absorption drops while E-layer and F-layer propagation remain strong during twilight hours.
- Gray-line propagation is long-distance communications at twilight on frequencies less than 15 MHz.

Group E3C – Radio Wave Propagation

Questions in group E3C cover the following topics: aurora propagation, selective fading, radio-path horizon, take-off angle over flat or sloping terrain, effects of ground on propagation, and less common propagation modes.

E3C Test Notes

- SSB signals are raspy, CW signals appear to be modulated by white noise, and signals propagating through the aurora are fluttery during periods of aurora activity.

- Aurora activity is caused by the interaction of charged particles from the sun with Earth's magnetic field and the ionosphere.

- Aurora activity occurs in the E-region of the ionosphere.

- CW is the best emission mode for aurora propagation.

- Selective fading is partial cancellation of some frequencies within the received passband.

- The VHF/UHF radio-path horizon distance exceeds the geometric horizon by approximately 15%.

- The main lobe take-off angle of a horizontally polarized three-element beam antenna decreases with increasing height above the ground.

- Pedersen ray is the high-angle wave in HF propagation that travels for some distance within the F2 region.

- Tropospheric ducting is usually responsible for causing VHF signals to propagate for hundreds of miles.

- Unlike the same antenna mounted on flat ground, the main lobe take-off angle of a horizontally polarized antenna mounted on the side of a hill decreases in the downhill direction.

- From the contiguous 48 states, an antenna should be pointed approximately north to take maximum advantage of aurora propagation.

- When signal frequency is increased, the maximum distance of ground-wave propagation decreases.

- Vertical polarization is best for ground-wave propagation.

- The radio-path horizon distance exceeds the geometric horizon because of downward bending due to density variations in the atmosphere.

5.4 SUBELEMENT E4

Subelement E4 contains five targeted question groups with a total of 70 potential test questions. Five of the 50 [Element 4] test questions will be drawn from subelement E4.

Group E4A – Test Equipment

Questions in group E4A cover the following topics: analog and digital instruments, spectrum and network analyzers, antenna analyzers, oscilloscopes, testing transistors, and RF measurements.

E4A Test Notes

- A spectrum analyzer displays signals in the frequency domain, while an oscilloscope displays signals in the time domain.
- Spectrum analyzers display frequency on the horizontal axis.
- Spectrum analyzers display amplitude on the vertical axis.
- A spectrum analyzer is used to display spurious signals from a radio transmitter.
- A spectrum analyzer is also used to display intermodulation distortion products in an SSB transmission.
- A spectrum analyzer could determine the degree of isolation between the input and output ports of a 2 meter duplexer, whether a crystal is operating on its fundamental or overtone frequency, and the spectral output of a transmitter.
- An advantage of using an antenna analyzer instead of an SWR bridge to measure antenna SWR is that antenna analyzers do not need an external RF source.
- An antenna analyzer is the best instrument for measuring the SWR of a beam antenna.

- A good method for measuring the intermodulation distortion of your own PSK signal is to transmit into a dummy load, receive the signal on a second receiver, and feed the audio into the sound card of a computer running an appropriate PSK program.
- To establish if a silicon NPN junction transistor is biased on, measure base-to-emitter voltage with a voltmeter; it should be approximately 0.6 to 0.7 volts.
- An oscilloscope can be used for detailed analysis of digital signals.
- It is important to attenuate the transmitter output when connecting to a spectrum analyzer.

Group E4B – Measurement Technique and Limitations

Questions in group E4B cover the following topics: instrument accuracy and performance limitations, probes, techniques to minimize errors, measurement of "Q", and instrument calibration.

E4B Test Notes

- Time base accuracy most affects the accuracy of a frequency counter.
- An advantage of using a bridge circuit to measure impedance is that the measurement is based on obtaining a signal null, which can be done very precisely.
- If a frequency counter with a specified accuracy of +/- 1.0 ppm reads 146,520,000 Hz, the actual frequency being measured could differ by 146.52 Hz at most.
- If a frequency counter with a specified accuracy of +/- 0.1 ppm reads 146,520,000 Hz, the actual frequency being measured could differ by 14.652 Hz at most.
- If a frequency counter with a specified accuracy of +/- 10 ppm reads 146,520,000 Hz, the actual frequency being measured could differ by 1465.20 Hz at most.
- When a directional power meter connected between a transmitter and a terminating load reads 100 watts forward power and 25 watts reflected power, 75 watts of power are absorbed by the load.

- It is considered good practice to keep the signal ground connection of an oscilloscope probe as short as possible.
- High impedance input is characteristic of a good DC voltmeter.
- If the current reading on an RF ammeter (placed in series with the antenna feed line of a transmitter) increases as the transmitter is tuned to resonance, it is an indication that there is more power going into the antenna.
- A method to measure intermodulation distortion in an SSB transmitter is to modulate the transmitter with two non-harmonically-related audio frequencies and observe the RF output with a spectrum analyzer.
- Connect the antenna feed line directly to a portable antenna analyzer's connector when measuring antenna resonance and feed-point impedance.
- The full scale reading of a voltmeter multiplied by its ohms-per-volt rating (voltmeter sensitivity) will provide the input impedance of the voltmeter.
- To calibrate the compensation of an oscilloscope probe, adjust the probe until the horizontal portions of the displayed square wave are as nearly flat as possible.
- A dip meter too tightly coupled to a tuned circuit could result in a less accurate reading.
- The bandwidth of the circuit's frequency response can be used as a relative measurement of the Q for a series-tuned circuit.

Group E4C – Receiver Performance Characteristics

Questions in group E4C cover the following topics: phase noise, capture effect, noise floor, image rejection, MDS, signal-to-noise-ratio, and selectivity.

E4C Test Notes

- Phase noise in the local oscillator section of a receiver can cause strong signals on nearby frequencies to interfere with reception of weak signals.
- The front-end filter or pre-selector portion of a receiver can be effective in eliminating image signal interference.
- Capture effect is the blocking of one FM phone signal by another stronger FM phone signal.

- The noise figure of a receiver is the ratio in dB of the noise generated by the receiver compared to the theoretical minimum noise.
- With regard to the noise floor of a receiver, the theoretical noise at the input of a perfect receiver at room temperature is -174 dBm/Hz.
- If a CW receiver with the AGC off has an equivalent input noise power density of -174 dBm/Hz, a -148 dBm unmodulated carrier input will yield an audio output SNR of 0 dB in a 400 Hz noise bandwidth.
- The MDS of a receiver is the minimum discernible signal.
- Lowering the noise figure can improve weak signal sensitivity in a receiver.
- Selecting a high frequency for design of the IF in a conventional HF or VHF communications receiver makes it easier for front-end circuitry to eliminate image responses.
- 300 Hz is a desirable amount of selectivity for an amateur RTTY HF receiver.
- 2.4 kHz is a desirable amount of selectivity for an amateur SSB phone receiver.
- Using too wide a filter bandwidth in a receiver's IF section may result in undesired signals.
- A narrow-band roofing filter improves the dynamic range of a receiver by attenuating strong signals near the receive frequency.
- A signal transmitting on 15.210 MHz can generate a spurious image signal in a receiver that is tuned to 14.300 MHz and using a 455 kHz IF frequency.
- Atmospheric noise is the primary source of noise heard from an HF receiver with an antenna connected.

Group E4D – Receiver Performance Characteristics

Questions in group E4D cover the following topics: blocking dynamic range, intermodulation and cross-modulation interference, third-order intercept, desensitization, and preselection.

E4D Test Notes

- The blocking dynamic range of a receiver is the difference in dB between the noise floor and the level of an incoming signal that will cause 1 dB of gain compression.
- Poor dynamic range in a communications receiver causes cross-modulation of the desired signal and desensitization from strong adjacent signals.
- Intermodulation interference occurs between two repeaters when the repeaters are in close proximity and the signals mix in the final amplifier of one or both transmitters.
- A properly terminated circulator at the output of the transmitter may reduce or eliminate intermodulation interference in a repeater caused by another transmitter operating in close proximity.
- Transmitter frequencies 146.34 MHz and 146.61 MHz can cause an intermodulation-product signal in a receiver tuned to 146.70 MHz when a nearby station transmits on 146.52 MHz.
- The term "intermodulation interference" describes unwanted signals generated by the mixing of two or more signals.
- Cross-modulation is when an unwanted off-frequency signal is heard in addition to the desired signal.
- Nonlinear circuits or devices cause intermodulation in an electronic circuit.
- The purpose of the preselector in a communications receiver is to increase rejection of unwanted signals.
- With respect to receiver performance, a third-order intercept level of 40 dBm means a pair of 40 dBm signals will theoretically generate a third-order intermodulation product with the same level as the input signals.
- Third-order intermodulation products created within a receiver are of particular interest compared to other products because the third-order product of two signals that are in the band of interest is also likely to be within the band.

- Desensitization is the reduction in receiver sensitivity caused by a strong signal near the received frequency.
- Strong adjacent-channel signals can cause receiver desensitization.
- To reduce the likelihood of receiver desensitization, decrease the receiver RF bandwidth.

Group E4E – Noise Suppression

Questions in group E4E cover the following topics: system noise, electrical appliance noise, line noise, locating noise sources, DSP noise reduction, and noise blankers.

E4E Test Notes

- Ignition noise can often be reduced by use of a receiver noise blanker.
- Broadband white noise, ignition noise, and power line noise can often be reduced with a DSP noise filter.
- A receiver noise blanker might be able to remove signals that appear across a wide bandwidth.
- Conducted and radiated noise caused by an automobile alternator can be suppressed by connecting the radio's power leads directly to the battery and by installing coaxial capacitors in line with the alternator leads.
- Noise from an electric motor can be suppressed by installing a brute-force AC-line filter in series with the motor leads.
- Thunderstorms are a major cause of atmospheric static.
- You can determine if line noise interference is being generated within your home by turning off the AC power line main circuit breaker and listening on a battery-operated radio.
- Electrical wiring near a radio antenna picks up a common-mode signal at the frequency of the radio transmitter.
- Nearby signals may appear to be excessively wide even if they meet emission standards when using an IF noise blanker.
- Characteristics of interference caused by a touch-controlled electrical device include the following: interfering signals sound like AC hum on an AM receiver or a carrier modulated by 60 Hz hum on a SSB or CW

receiver, interfering signals may drift slowly across the HF spectrum, and interfering signals can be several kHz in width and usually repeat at regular intervals across an HF band.

- Hearing combinations of local AM broadcast signals within one or more of the MF or HF ham bands is likely caused by nearby corroded metal joints mixing and re-radiating the broadcast signals.

- When attempting to copy CW signals, some types of automatic DSP notch-filters can remove the desired signal at the same time as it removes interfering signals.

- A loud roaring or buzzing AC line interference that comes and goes at intervals can be caused by arcing contacts in a thermostatically controlled device, a defective doorbell or doorbell transformer inside a nearby residence, or a malfunctioning illuminated advertising display.

- Operation of a nearby personal computer can cause the appearance of unstable modulated or unmodulated signals at specific frequencies.

5.5 SUBELEMENT E5

Subelement E5 contains four targeted question groups with a total of 71 potential test questions. Four of the 50 [Element 4] test questions will be drawn from subelement E5.

Group E5A – Resonance and Q

Questions in group E5A cover the following topics: characteristics of resonant circuits, series and parallel resonance, Q, half-power bandwidth, and phase relationships in reactive circuits.

E5A Test Notes

- Resonance can cause the voltage across reactances in series to be larger than the voltage applied to them.

- Resonance in an electrical circuit is the frequency at which the capacitive reactance equals the inductive reactance.

- The magnitude of the impedance of a series RLC circuit at resonance is approximately equal to circuit resistance.
- The magnitude of the impedance of a circuit with a resistor, an inductor, and a capacitor all in parallel at resonance is approximately equal to circuit resistance.
- Maximum current circulates at the input of a series RLC circuit at resonance.
- Maximum current circulates within the components of a parallel LC circuit at resonance.
- Minimum current circulates at the input of a parallel RLC circuit at resonance.
- The current through and the voltage across a series resonant circuit at resonance are in phase.
- The half-power bandwidth of a parallel resonant circuit that has a resonant frequency of 1.8 MHz and a Q of 95 is 18.9 kHz.
- The half-power bandwidth of a parallel resonant circuit that has a resonant frequency of 7.1 MHz and a Q of 150 is 47.3 kHz.
- The half-power bandwidth of a parallel resonant circuit that has a resonant frequency of 3.7 MHz and a Q of 118 is 31.4 kHz.
- The half-power bandwidth of a parallel resonant circuit that has a resonant frequency of 14.25 MHz and a Q of 187 is 76.2 kHz.
- The resonant frequency of a series RLC circuit when R is 22 ohms, L is 50 microhenrys, and C is 40 picofarads, equals 3.56 MHz.
- The resonant frequency of a series RLC circuit when R is 56 ohms, L is 40 microhenrys, and C is 200 picofarads, equals 1.78 MHz.
- The resonant frequency of a parallel RLC circuit when R is 33 ohms, L is 50 microhenrys, and C is 10 picofarads, equals 7.12 MHz.
- The resonant frequency of a parallel RLC circuit when R is 47 ohms, L is 25 microhenrys and C is 10 picofarads, equals 10.1 MHz.

Group E5B – Time Constants and Phase Relationships

Questions in group E5B cover the following topics: RLC time constants, definitions, time constants in RL and RC circuits, phase angle between voltage and current, and phase angles of series and parallel circuits.

E5B Test Notes

- One time constant is the time required for the capacitor in an RC circuit to charge to 63.2% of the applied voltage.

- One time constant is the time it takes for a charged capacitor in an RC circuit to discharge to 36.8% of its initial voltage.

- The capacitor in an RC circuit is discharged to 13.5% of the starting voltage after two time constants.

- The time constant of a circuit having two 220-microfarad capacitors and two 1-megohm resistors all in parallel is 220 seconds.

- It takes 0.02 seconds for an initial charge of 20 V DC to decrease to 7.36 V DC in a 0.01-microfarad capacitor, when a 2-megohm resistor is connected across it.

- It takes 450 seconds for an initial charge of 800 V DC to decrease to 294 V DC in a 450-microfarad capacitor, when a 1-megohm resistor is connected across it.

- If XC is 500 ohms, R is 1 kilohm, and XL is 250 ohms, phase angle of a series RLC circuit is 14.0 degrees with the voltage lagging the current.

- If XC is 100 ohms, R is 100 ohms, and XL is 75 ohms, phase angle of a series RLC circuit is 14 degrees with the voltage lagging the current.

- The current through a capacitor leads the voltage across a capacitor by 90 degrees.

- The voltage across an inductor leads the current through an inductor by 90 degrees.

- If XC is 25 ohms, R is 100 ohms, and XL is 50 ohms, phase angle of a series RLC circuit is 14 degrees with the voltage leading the current.

- If XC is 75 ohms, R is 100 ohms, and XL is 50 ohms, phase angle of a series RLC circuit is 14 degrees with the voltage lagging the current.

- If XC is 250 ohms, R is 1 kilohm, and XL is 500 ohms, phase angle of a series RLC circuit is 14.04 degrees with the voltage leading the current.

Group E5C – Impedance Plots and Coordinate Systems

Questions in group E5C cover the following topics: plotting impedances in polar coordinates, and rectangular coordinates. Several questions from this group are based on the figure shown below.

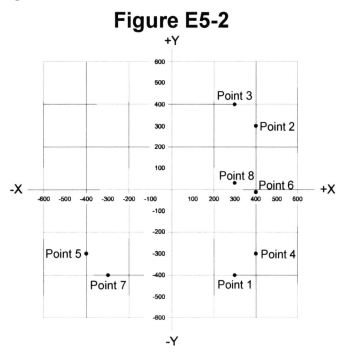

E5C Test Notes

- In polar coordinates, the impedance of a network consisting of a 100-ohm-reactance inductor in series with a 100-ohm resistor is 141 ohms at an angle of 45 degrees.
- In polar coordinates, the impedance of a network consisting of a 100-ohm-reactance inductor, a 100-ohm-reactance capacitor, and a 100-ohm resistor all connected in series is 100 ohms at an angle of 0 degrees.
- In polar coordinates, the impedance of a network consisting of a 300-ohm-reactance capacitor, a 600-ohm-reactance inductor, and a 400-ohm resistor all connected in series is 500 ohms at an angle of 37 degrees.

- In polar coordinates, the impedance of a network consisting of a 400-ohm-reactance capacitor in series with a 300-ohm resistor is 500 ohms at an angle of -53.1 degrees.
- In polar coordinates, the impedance of a network consisting of a 400-ohm-reactance inductor in parallel with a 300-ohm resistor is 240 ohms at an angle of 36.9 degrees.
- In polar coordinates, the impedance of a network consisting of a 100-ohm-reactance capacitor in series with a 100-ohm resistor is 141 ohms at an angle of -45 degrees.
- In polar coordinates, the impedance of a network comprised of a 100-ohm-reactance capacitor in parallel with a 100-ohm resistor is 71 ohms at an angle of -45 degrees.
- In polar coordinates, the impedance of a network comprised of a 300-ohm-reactance inductor in series with a 400-ohm resistor is 500 ohms at an angle of 37 degrees.
- When using rectangular coordinates to graph the impedance of a circuit, the horizontal axis represents resistive components.
- When using rectangular coordinates to graph the impedance of a circuit, the vertical axis represents reactive components.
- The two numbers that define a point on a graph using rectangular coordinates represent the coordinate values along the horizontal and vertical axes.
- If you plot the impedance of a circuit using the rectangular coordinate system and find the impedance point falls on the right side of the graph on the horizontal axis, the circuit is equivalent to a pure resistance.
- Rectangular coordinates are often used to display the resistive, inductive, and/or capacitive reactance components of an impedance.
- Polar coordinates are often used to display the phase angle of a circuit containing resistance, inductive and/or capacitive reactance.
- In polar coordinates, the impedance of a circuit of 100 -j100 ohms impedance is 141 ohms at an angle of -45 degrees.

- In polar coordinates, the impedance of a circuit that has an admittance of 7.09 millisiemens at 45 degrees is 141 ohms at an angle of -45 degrees.
- In rectangular coordinates, the impedance of a circuit that has an admittance of 5 millisiemens at -30 degrees is 173 +j100 ohms.
- In polar coordinates, the impedance of a series circuit consisting of a resistance of 4 ohms, an inductive reactance of 4 ohms, and a capacitive reactance of 1 ohm is 5 ohms at an angle of 37 degrees
- Point 4 in Figure E5-2 represents the impedance of a series circuit consisting of a 400 ohm resistor and a 38 picofarad capacitor at 14 MHz.
- Point 3 in Figure E5-2 represents the impedance of a series circuit consisting of a 300 ohm resistor and an 18 microhenry inductor at 3.505 MHz.
- Point 1 in Figure E5-2 represents the impedance of a series circuit consisting of a 300 ohm resistor and a 19 picofarad capacitor at 21.200 MHz.
- In rectangular coordinates, the impedance of a network consisting of a 10-microhenry inductor in series with a 40-ohm resistor at 500 MHz is 40 + j31,400.
- Point 8 in Figure E5-2 represents the impedance of a series circuit consisting of a 300-ohm resistor, a 0.64-microhenry inductor, and an 85-picofarad capacitor at 24.900 MHz.

Group E5D – AC and RF Energy in Real Circuits

Questions in group E5D cover the following topics: skin effect, electrostatic and electromagnetic fields, reactive power, power factor, and coordinate systems.

E5D Test Notes
- Skin effect occurs as frequency increases; RF current flows in a thinner layer of the conductor, closer to the surface.
- The resistance of a conductor is different for RF currents than for direct currents because of skin effect.
- A capacitor is used to store electrical energy in an electrostatic field.
- Joules measure electrical energy stored in an electrostatic field.

- Electric current creates a magnetic field.
- You can determine what direction a magnetic field is oriented about a conductor in relation to the direction of electron flow by using the left-hand rule.
- The amount of current determines the strength of a magnetic field around a conductor.
- Potential energy is stored in an electromagnetic or electrostatic field.
- Reactive power in an AC circuit that has both ideal inductors and ideal capacitors is repeatedly exchanged between the associated magnetic and electric fields but is not dissipated.
- The true power can be determined in an AC circuit where the voltage and current are out of phase by multiplying the apparent power times the power factor.
- An R-L circuit with a 60-degree phase angle between the voltage and the current has a power factor of 0.5.
- 80 watts of power is consumed in a circuit having a power factor of 0.2, if the input is 100V AC at 4 amperes.
- 100 watts of power is consumed in a circuit consisting of a 100 ohm resistor in series with a 100 ohm inductive reactance and drawing 1 ampere.
- Reactive power is wattless, nonproductive power.
- An RL circuit with a 45-degree phase angle between the voltage and the current has a power factor of 0.707.
- An RL circuit with a 30-degree phase angle between the voltage and the current has a power factor of 0.866.
- 600 watts of power are consumed in a circuit having a power factor of 0.6, if the input is 200V AC at 5 amperes.
- 355 watts of power are consumed in a circuit having a power factor of 0.71, if the apparent power is 500 watts.

5.6 SUBELEMENT E6

Subelement E6 contains six targeted question groups with a total of 83 potential test questions. Six of the 50 [Element 4] test questions will be drawn from subelement E6.

Group E6A – Semiconductor Materials and Devices

Questions in group E6A cover the following topics: germanium, silicon, P-type, N-type, transistor types, NPN, PNP, junction, field-effect transistors, enhancement mode, depletion mode, MOS, CMOS, N-channel, and P-channel. Some questions from this group are based on the figures shown below.

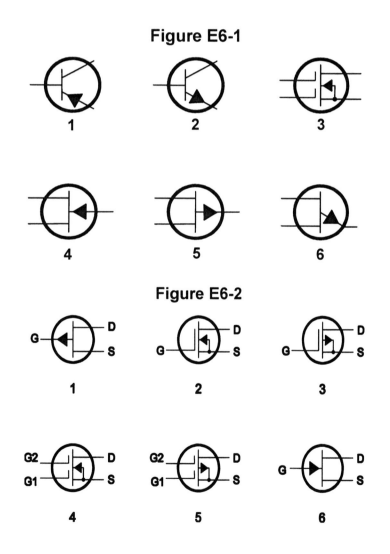

E6A Test Notes

- Gallium arsenide is preferred as a semiconductor material over germanium or silicon at microwave frequencies.
- N-type semiconductor materials contain an excess of free electrons.
- The majority charge carriers in P-type semiconductor material are holes.
- An impurity atom that adds holes to a semiconductor crystal structure is called an acceptor impurity.
- The alpha of a bipolar junction transistor is the change of collector current with respect to emitter current.
- The beta of a bipolar junction transistor is the change in collector current with respect to base current.
- Schematic symbol 1 in Figure E6-1 represents a PNP transistor.
- Alpha cutoff frequency is the frequency at which the grounded-base current gain of a transistor has decreased to 0.7 of the gain obtainable at 1 kHz.
- A depletion-mode FET is an FET that exhibits a current flow between source and drain when no gate voltage is applied.
- Schematic symbol 4 in Figure E6-2 represents an N-channel dual-gate MOSFET.
- Schematic symbol 1 in Figure E6-2 represents a P-channel junction FET.
- Many MOSFET devices have internally connected Zener diodes on the gates to reduce the chance of the gate insulation being punctured by static discharges or excessive voltages.
- CMOS stands for complementary metal-oxide semiconductor.
- An FET has high input impedance, and a bipolar transistor has low input impedance.
- P-type semiconductor materials contain an excess of holes in the outer shell of electrons.
- The majority charge carriers in N-type semiconductor material are free electrons.

- The three terminals of a field-effect transistor are gate, drain, and source.

Group E6B – Circuit Components

Questions in group E6B cover semiconductor diodes. Some questions from this group are based on the figure shown below.

Figure E6-3

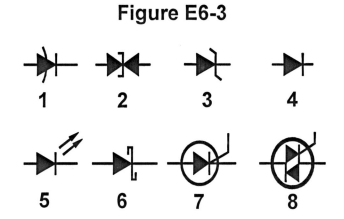

E6B Test Notes

- The most useful characteristic of a Zener diode is a constant voltage drop under conditions of varying current.
- A Schottky diode has less forward voltage drop compared to an ordinary silicon diode.
- A tunnel diode is capable of both amplification and oscillation.
- A varactor diode is designed for use as a voltage-controlled capacitor.
- The large region of intrinsic material of a PIN diode makes it useful as an RF switch or attenuator.
- Hot-carrier diodes are commonly used as VHF / UHF mixers or detectors.
- Excessive junction temperature will cause a junction diode to fail when supplied too much current.
- A metal-semiconductor junction is a type of semiconductor diode.
- Point contact diodes are commonly used as RF detectors.
- Schematic symbol 5 in Figure E6-3 represents a light-emitting diode (LED).
- Forward DC bias current is used to control the attenuation of RF signals by a PIN diode.

- PIN diodes are commonly used as RF switches.
- Forward bias is required for an LED to emit light.

Group E6C – Integrated Circuits

Questions in group E6C cover TTL digital integrated circuits, CMOS digital integrated circuits, plus gates. Some questions from this group are based on the figure shown below.

Figure E6-5

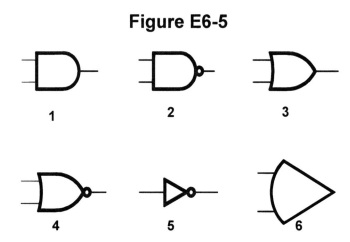

E6C Test Notes

- The recommended power supply voltage for TTL series integrated circuits is 5 volts.
- The inputs of a TTL device assume a logic-high state if they are left open.
- Tri-state logic devices have 0, 1, and high impedance output states.
- The primary advantage of tri-state logic is the ability to connect many device outputs to a common bus.
- An advantage of CMOS logic devices over TTL devices is lower power consumption.
- CMOS digital integrated circuits have high immunity to noise on the input signal or power supply because the input switching threshold is about one-half the power supply voltage.
- Schematic symbol 1 in Figure E6-5 represents an AND gate.
- Schematic symbol 2 in Figure E6-5 represents a NAND gate.
- Schematic symbol 3 in Figure E6-5 represents an OR gate.
- Schematic symbol 4 in Figure E6-5 represents a NOR gate.

- Schematic symbol 5 in Figure E6-5 represents the NOT operation (inverter).
- BiCMOS logic is an integrated circuit logic family using both bipolar and CMOS transistors.
- An advantage of BiCMOS logic is that it has the high input impedance of CMOS and the low output impedance of bipolar transistors.

Group E6D – Optical Devices and Toroids

Questions in group E6D cover the following topics: cathode-ray tube devices, charge-coupled devices (CCDs), liquid crystal displays (LCDs), toroids, permeability, core material, selecting, and winding.

E6D Test Notes

- Cathode-ray tube (CRT) persistence is the length of time the image remains on the screen after the beam is turned off.
- Exceeding the anode voltage rating can cause a cathode-ray tube (CRT) to generate X-rays.
- A charge-coupled device (CCD) samples an analog signal and passes it in stages from the input to the output.
- A charge-coupled device (CCD) stores photogenerated charges as signals corresponding to pixels in modern video cameras.
- A liquid crystal display (LCD) is a display using a crystalline liquid, which in conjunction with polarizing filters becomes opaque when voltage is applied.
- Permeability determines the inductance of a toroidal inductor.
- Assuming a correct selection of core material for the frequency being used, the usable frequency range of inductors that use toroidal cores is from less than 20 Hz to approximately 300 MHz.
- One important reason for using powdered-iron toroids rather than ferrite toroids in an inductor is because powdered-iron toroids generally maintain their characteristics at higher currents.
- Ferrite beads are commonly used as VHF and UHF parasitic suppressors at the input and output terminals of transistorized HF amplifiers.

- A primary advantage of using a toroidal core instead of a solenoidal core in an inductor is that toroidal cores confine most of the magnetic field within the core material.
- 43 turns are required to produce a 1-mH inductor using a ferrite toroidal core that has an inductance index (A L) value of 523 millihenrys/1000 turns.
- 35 turns are required to produce a 5-microhenry inductor using a powdered-iron toroidal core that has an inductance index (A L) value of 40 microhenrys/100 turns.
- Electrostatic CRT deflection is better when high-frequency waveforms are to be displayed on the screen.
- Charge-coupled devices (CCDs) use a combination of analog and digital circuitry, can be used to make audio delay lines, and sample and store analog signals.
- The main advantage of liquid crystal display (LCD) devices over other types of display devices is that they consume less power.
- One reason for using ferrite toroids rather than powdered-iron toroids in an inductor is because ferrite toroids generally require fewer turns to produce a given inductance value.

Group E6E – Piezoelectric Crystals and MMICs

Questions in group E6E cover quartz crystals, crystal oscillators and filters, plus monolithic amplifiers.

E6E Test Notes
- A crystal lattice filter is a filter with narrow bandwidth and steep skirts made using quartz crystals.
- The relative frequencies of individual crystals have the greatest effect in helping determine bandwidth and response shape of a crystal ladder filter.
- One aspect of the piezoelectric effect is physical deformation of a crystal by the application of a voltage.
- The most common input and output impedance of circuits that use MMICs is 50 ohms.

- A 2 dB noise figure value is typical of a low-noise UHF preamplifier.
- Controlled gain, low noise figure, and constant input and output impedance over the specified frequency range make the MMIC a popular choice for VHF through microwave circuits.
- Microstrip construction is typically used to construct an MMIC-based microwave amplifier.
- Power-supply voltage is normally furnished to the most common type of monolithic microwave integrated circuit (MMIC) through a resistor and/or RF choke connected to the amplifier output lead.
- To ensure a crystal oscillator provides the frequency specified by the crystal manufacturer, you must provide the crystal with a specified parallel capacitance.
- The equivalent circuit of a quartz crystal is motional capacitance, motional inductance and loss resistance in series, with a shunt capacitance representing electrode and stray capacitance.
- Gallium nitride is likely to provide the highest frequency of operation when used in MMICs.
- A Jones filter is used as a variable bandwidth crystal lattice filter in an HF receiver IF stage.

Group E6F – Optical Components and Power Systems

Questions in group E6F cover the following topics: photoconductive principles and effects, photovoltaic systems, optical couplers, optical sensors, and optoisolators.

E6F Test Notes

- Photoconductivity is the increased conductivity of an illuminated semiconductor.
- When light shines on photoconductive material, its conductivity increases.
- The most common configuration of an optoisolator or optocoupler is an LED and a phototransistor.
- The photovoltaic effect is the conversion of light to electrical energy.

- An optical shaft encoder is a device that detects rotation of a control by interrupting a light source with a patterned wheel.
- Crystalline semiconductors are affected the most by photoconductivity.
- A solid-state relay is a device that uses semiconductor devices to implement the functions of an electromechanical relay.
- Optoisolators are often used in conjunction with solid-state circuits when switching 120V AC because they provide a very high degree of electrical isolation between a control circuit and the circuit being switched.
- Efficiency of a photovoltaic cell is the relative fraction of light that is converted to current.
- Silicon is the most common type of photovoltaic cell used for electrical power generation.
- The open-circuit voltage produced by a fully illuminated silicon photovoltaic cell is approximately 0.5 volts.
- Electrons absorb the energy from light falling on a photovoltaic cell.

5.7 SUBELEMENT E7

Subelement E7 contains eight targeted question groups with a total of 125 potential test questions. Eight of the 50 [Element 4] test questions will be drawn from subelement E7.

Group E7A – Digital Circuits

Questions in group E7A cover the following topics: digital circuit principles and logic circuits, classes of logic elements, positive and negative logic, frequency dividers, and truth tables.

E7A Test Notes

- A flip-flop is a bistable circuit.
- Two output level changes are obtained for every two trigger pulses applied to the input of a T flip-flop circuit.
- A flip-flop can divide the frequency of a pulse train by 2.

- Two flip-flops are required to divide a signal frequency by 4.
- Astable multivibrators continuously alternate between two states without an external clock.
- A monostable multivibrator switches momentarily to the opposite binary state and then returns, after a set time, to its original state.
- A NAND gate produces a logic "0" at its output only when all inputs are logic "1".
- An OR gate produces a logic "1" at its output if any or all inputs are logic "1".
- A two-input exclusive NOR gate produces a logic "0" at its output if any single input is a logic "1".
- A truth table is a list of inputs and corresponding outputs for a digital device.
- Positive logic represents a logic "1" as a high voltage.
- Negative logic represents a logic "0" as a high voltage.
- An SR or RS flip-flop is a set/reset flip-flop whose output is low when R is high and S is low, high when S is high and R is low, and unchanged when both inputs are low.
- A JK flip-flop is similar to an RS flip-flop except that it toggles when both J and K are high.
- A D flip-flop is a flip-flop whose output takes on the state of the D input when the clock signal transitions from low to high.

Group E7B – Amplifiers

Questions in group E7B cover the following topics: class of operation, vacuum tube and solid-state circuits, distortion and intermodulation, spurious and parasitic suppression, plus microwave amplifiers. Some questions from this group are based on the figures shown below.

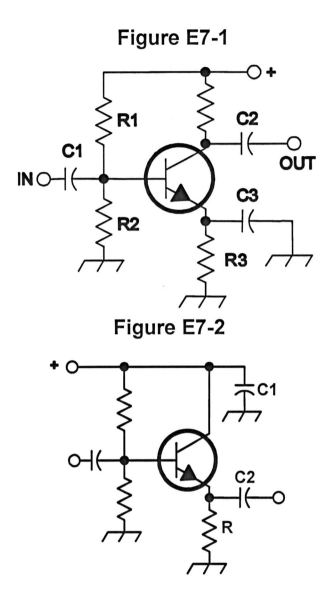

Figure E7-1

Figure E7-2

E7B Test Notes

- Class AB amplifiers operate more than 180 degrees but less than 360 degrees of a signal cycle.
- A Class D amplifier is a type of amplifier that uses switching technology to achieve high efficiency.
- The output of a class D amplifier circuit is a low-pass filter to remove switching signal components.
- Bias on the load line of a Class A common emitter amplifier is usually set approximately halfway between saturation and cutoff.
- To prevent unwanted oscillations in an RF power amplifier, install parasitic suppressors and/or neutralize the stage.

- Push-pull amplifiers reduce or eliminate even-order harmonics.
- Signal distortion and excessive bandwidth are likely when a Class C amplifier is used to amplify a single-sideband phone signal.
- An RF power amplifier can be neutralized by feeding a 180-degree out-of-phase portion of the output back to the input.
- When tuning a vacuum tube RF power amplifier that employs a pi network output circuit, the tuning capacitor is adjusted for minimum plate current and the loading capacitor is adjusted for maximum permissible plate current.
- In Figure E7-1, R1 and R2 provide fixed bias.
- In Figure E7-1, R3 provides self bias.
- The circuit shown in Figure E7-1 is a common emitter amplifier.
- In Figure E7-2, R provides an emitter load.
- In Figure E7-2, C2 provides output coupling.
- One way to prevent thermal runaway in a bipolar transistor amplifier is to use a resistor in series with the emitter.
- Intermodulation products in a linear power amplifier result in transmission of spurious signals.
- Third-order intermodulation distortion products are of particular concern in linear power amplifiers because they are relatively close in frequency to the desired signal.
- A grounded-grid amplifier has low input impedance.
- A klystron is a VHF, UHF, or microwave vacuum tube that uses velocity modulation.
- A parametric amplifier is a low-noise VHF or UHF amplifier relying on varying reactance for amplification.
- Field-effect transistors are generally best suited for UHF or microwave power amplifier applications.

Group E7C – Filters and Impedance Matching Networks

Questions in group E7C cover the following topics: types of networks, types of filters, filter applications, filter characteristics, impedance matching, and DSP filtering.

E7C Test Notes

- In a low-pass filter pi network, a capacitor is connected between the input and ground, another capacitor is connected between the output and ground, and an inductor is connected between input and output.
- A T-network with series capacitors and a parallel shunt inductor is a high-pass filter.
- For impedance matching between the final amplifier of a vacuum tube transmitter and an antenna, a pi-L-network has greater harmonic suppression than a pi network.
- An impedance matching circuit transforms a complex impedance to a resistive impedance by cancelling the reactive part of the impedance and changing the resistive part to a desired value.
- A Chebyshev filter has ripple in the passband and a sharp cutoff.
- An elliptical filter has extremely sharp cutoff with one or more notches in the stop band.
- Use a notch filter to attenuate an interfering carrier signal while receiving an SSB transmission.
- An adaptive filter is a digital signal processing audio filter that can be used to remove unwanted noise from a received SSB signal.
- A Hilbert transform filter is a digital signal processing filter that can be used to generate an SSB signal.
- A cavity filter is the best choice for use in a 2 meter repeater duplexer.
- A pi network is equivalent to two L networks connected back-to-back with the inductors in series and the capacitors in shunt at the input and output.
- A pi-L network used for matching a vacuum tube final amplifier to a 50-ohm unbalanced output can be described as a pi network with an additional series inductor on the output.

- An advantage of a pi matching network over an L matching network consisting of a single inductor and a single capacitor is that the Q of a pi network can be varied depending on the component values chosen.
- Digital mode is most affected by non-linear phase response in a receiver IF filter.

Group E7D – Practical Circuits

Questions in group E7D cover power supplies and voltage regulators. Several questions from this group are based on the figure shown below.

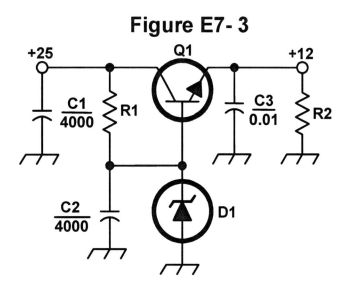

Figure E7-3

E7D Test Notes

- The conduction of a control element in a linear electronic voltage regulator is varied to maintain a constant output voltage.
- The control device's duty cycle in a switching electronic voltage regulator is controlled to produce a constant average output voltage.
- A Zener diode is typically used as a stable reference voltage in a linear voltage regulator.
- A series regulator usually makes the most efficient use of the primary power source.
- A shunt regulator places a constant load on the unregulated voltage source.
- Q1 in Figure E7-3 increases the current-handling capability of the regulator.

- C2 in Figure E7-3 bypasses hum around D1.
- The circuit shown in Figure E7-3 is a linear voltage regulator.
- C1 in Figure E7-3 filters the supply voltage.
- C3 in Figure E7-3 prevents self-oscillation.
- R1 in Figure E7-3 supplies current to D1.
- R2 in Figure E7-3 provides a constant minimum load for Q1.
- D1 in Figure E7-3 provides a voltage reference.
- One purpose of a bleeder resistor in a conventional (unregulated) power supply is to improve output voltage regulation.
- A step-start circuit in a high-voltage power supply allows the filter capacitors to charge gradually.
- When several electrolytic filter capacitors are connected in series to increase the operating voltage of a power supply filter circuit, resistors should be connected across each capacitor to equalize the voltage drop across each capacitor as much as possible, to provide a safety bleeder to discharge the capacitors when the supply is off, and to provide a minimum load current to reduce voltage excursions at light loads.
- A high-frequency inverter type high-voltage power supply can be less expensive and lighter than a conventional power supply because the high frequency inverter design uses much smaller transformers and filter components for an equivalent power output.

Group E7E – Modulation and Demodulation

Questions in group E7E cover the following topics: reactance, phase and balanced modulators, detectors, mixer stages, DSP modulation and demodulation, and software-defined radio systems.

E7E Test Notes

- A reactance modulator on the oscillator can be used to generate FM phone emissions.
- A reactance modulator produces PM signals by using an electrically variable inductance or capacitance.

- An analog phase modulator varies the tuning of an amplifier tank circuit to produce PM signals.
- A single-sideband phone signal can be generated by using a balanced modulator followed by a filter.
- A pre-emphasis network circuit is added to an FM transmitter to boost the higher audio frequencies.
- De-emphasis is commonly used in FM communications receivers for compatibility with transmitters using phase modulation.
- The term "baseband" refers to the frequency components present in a modulating signal.
- The two input frequencies, along with their sum and difference frequencies, appear at the output of a mixer circuit.
- When an excessive amount of signal energy reaches a mixer circuit, spurious mixer products are generated.
- A diode detector functions by the rectification and filtering of RF signals.
- A product detector is well suited for demodulating SSB signals.
- A frequency discriminator stage in an FM receiver is a circuit for detecting FM signals.
- The quadrature method is a common means of generating an SSB signal when using digital signal processing.
- When referring to a software-defined receiver, direct conversion occurs when incoming RF is mixed to "baseband" for analog-to-digital conversion and subsequent processing.

Group E7F – Frequency Markers and Counters

Questions in group E7F cover frequency divider circuits, frequency marker generators, and frequency counters.

E7F Test Notes

- A prescaler circuit divides a higher frequency signal so a low-frequency counter can display the input frequency.
- A prescaler can be used to reduce a signal's frequency by a factor of ten.

- A decade counter digital IC produces one output pulse for every ten input pulses.
- Two flip-flops must be added to a 100-kHz crystal-controlled marker generator to provide markers at 50 and 25 kHz.
- Using a GPS signal reference, a rubidium-stabilized reference oscillator, and a temperature-controlled high Q dielectric resonator are all techniques for providing the high stability oscillators needed for microwave transmission and reception.
- One purpose of a marker generator is to provide a means of calibrating a receiver's frequency settings.
- The accuracy of the time base determines the accuracy of a frequency counter.
- A frequency counter counts the number of input pulses occurring within a specific period of time.
- A frequency counter provides a digital representation of the frequency of a signal.
- Instead of directly counting input pulses, some counters use period measurement plus mathematical computation to determine frequency.
- An advantage of a period-measuring frequency counter over a direct-count type is that it provides improved resolution of low-frequency signals within a comparable time period.

Group E7G – Active Filters and Op-amps

Questions in group E7G cover the following topics: active audio filters, characteristics, basic circuit design, and operational amplifiers. Some questions from this group are based on the figure shown below.

Figure E7-4

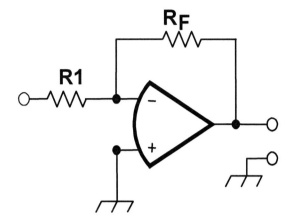

E7G Test Notes

- The values of capacitors and resistors external to the op-amp determine the gain and frequency characteristics of an op-amp RC active filter.
- Ringing in a filter adds unwanted oscillations to the desired signal.
- An advantage of using op-amps instead of LC elements in audio filters is that op-amps exhibit gain rather than insertion loss.
- Polystyrene is a type of capacitor best suited for use in high-stability op-amp RC active filter circuits.
- Unwanted ringing and audio instability can be prevented in a multi-section op-amp RC audio filter circuit by restricting both gain and Q.
- The most appropriate use of an op-amp active filter is as an audio filter in a receiver.
- When R1 is 10 ohms and RF is 470 ohms, the expected voltage gain from the circuit in Figure E7-4 is 47.
- The gain of an ideal operational amplifier does not vary with frequency.
- If R1 is 1000 ohms, RF is 10,000 ohms, and 0.23 volts DC is applied to the input of the circuit shown in Figure E7-4, the output voltage will be -2.3 volts.
- When R1 is 1800 ohms and RF is 68 kilohms, the absolute voltage gain from the circuit in Figure E7-4 is 38.

- When R1 is 3300 ohms and RF is 47 kilohms, the absolute voltage gain from the circuit in Figure E7-4 is 14.
- An integrated circuit operational amplifier is a high-gain, direct-coupled differential amplifier with very high input and very low output impedance.
- Op-amp input-offset voltage is the differential input voltage needed to bring the open-loop output voltage to zero.
- The typical input impedance of an integrated circuit op-amp is very high.
- The typical output impedance of an integrated circuit op-amp is very low.

Group E7H – Oscillators and Signal Sources

Questions in group E7H cover the following topics: types of oscillators, synthesizers and phase-locked loops, plus direct digital synthesizers.

E7H Test Notes

- Colpitts, Hartley, and Pierce are three oscillator circuits used in amateur radio equipment.
- A circuit must have positive feedback with a gain greater than 1 to oscillate.
- Positive feedback is supplied in a Hartley oscillator through a tapped coil.
- Positive feedback is supplied in a Colpitts oscillator through a capacitive divider.
- Positive feedback is supplied in a Pierce oscillator through a quartz crystal.
- Colpitts and Hartley oscillator circuits are commonly used in VFOs.
- A magnetron oscillator is a UHF or microwave oscillator consisting of a diode vacuum tube with a specially shaped anode, surrounded by an external magnet.
- A Gunn diode oscillator is an oscillator based on the negative resistance properties of properly doped semiconductors.
- A direct digital synthesizer circuit uses a phase accumulator, lookup table, digital-to-analog converter, and a low-pass anti-alias filter.
- The amplitude values that represent a sine wave output are contained in the lookup table of a direct digital frequency synthesizer.

- Spurious signals at discrete frequencies are the major spectral impurity components of direct digital synthesizers.
- Phase accumulator is a principal component of a direct digital synthesizer (DDS).
- The capture range of a phase-locked loop circuit is the frequency range over which the circuit can lock.
- A phase-locked loop circuit is an electronic servo loop consisting of a phase detector, a low-pass filter, a voltage-controlled oscillator, and a stable reference oscillator.
- A phase-locked loop can perform frequency synthesis and FM demodulation.
- The short-term stability of the reference oscillator is important in the design of a phase-locked loop (PLL) frequency synthesizer because any phase variations in the reference oscillator signal will produce phase noise in the synthesizer output.
- A phase-locked loop is often used as part of a variable frequency synthesizer for receivers and transmitters because it makes it possible for a VFO to have the same degree of frequency stability as a crystal oscillator.
- Phase noise is the major spectral impurity component of a phase-locked loop synthesizer.

5.8 SUBELEMENT E8

Subelement E8 contains four targeted question groups with a total of 56 potential test questions. Four of the 50 [Element 4] test questions will be drawn from subelement E8.

Group E8A – AC Waveforms

Questions in group E8A cover the following topics: sine, square, sawtooth, and irregular waveforms; AC measurements; average and PEP of RF signals; plus pulse and digital signal waveforms.

E8A Test Notes

- A square wave is made up of a sine wave plus all of its odd harmonics.
- A sawtooth wave has a rise time significantly faster than its fall time (or vice versa).

- A sawtooth wave is made up of sine waves of a given fundamental frequency plus all its harmonics.
- The root-mean-square value of an AC voltage is equal to the DC voltage causing the same amount of heating in a resistor as the corresponding RMS AC voltage.
- The most accurate way of measuring the RMS voltage of a complex waveform is by measuring the heating effect in a known resistor.
- The ratio of PEP-to-average power in a typical single-sideband phone signal is approximately 2.5 to 1.
- The PEP-to-average power ratio of a single-sideband phone signal is determined by the characteristics of the modulating signal.
- The period of a wave is the time required to complete one cycle.
- Human speech produces an irregular waveform.
- Pulse waveforms exhibit narrow bursts of energy separated by periods of no signal.
- A pulse-modulated signal can be used for digital data transmission.
- Human speech, video signals, and data can all be conveyed using digital waveforms.
- An advantage of using digital signals instead of analog is that digital signals can be regenerated multiple times without error.
- Sequential sampling is commonly used to convert analog signals to digital signals.
- On a conventional oscilloscope, a stream of digital data bits looks like a series of pulses with varying patterns.

Group E8B – Modulation and Demodulation

Questions in group E8B cover the following topics: modulation methods, modulation index and deviation ratio, pulse modulation, plus frequency and time division multiplexing.

E8B Test Notes

- Modulation index is the ratio between the frequency deviation of an RF carrier wave, and the modulating frequency of its corresponding FM-phone signal.
- The modulation index of a phase-modulated emission does not depend on the RF carrier frequency.
- When the modulating frequency is 1 Hz, an FM-phone signal with a maximum frequency deviation of 3 kHz either side of the carrier frequency has a modulation index of 3.
- When modulated with a 2-kHz modulating frequency, an FM-phone signal with a maximum carrier deviation of plus or minus 6 kHz has a modulation index of 3.
- When the maximum modulation frequency is 3 kHz, an FM-phone signal with a maximum frequency swing of plus or minus 5 kHz has a deviation ratio of 1.67.
- When the maximum modulation frequency is 3.5 kHz, an FM-phone signal with a maximum frequency swing of plus or minus 7.5 kHz has a deviation ratio of 2.14.
- The transmitter's peak power in a pulse-width modulation system is greater than its average power because the signal duty cycle is less than 100%.
- The modulating signal in a pulse-position modulation system varies the time at which each pulse occurs.
- Deviation ratio is the ratio of the maximum carrier frequency deviation to the highest audio modulating frequency.
- Frequency division multiplexing can be used to combine several separate analog information streams into a single analog radio frequency signal.
- Frequency division multiplexing is when two or more information streams are merged into a "baseband," which then modulates the transmitter.
- Digital time division multiplexing is when two or more signals are arranged to share discrete time slots of a data transmission.

Group E8C – Digital Signals

Questions in group E8C cover the following topics: digital communications modes, CW, information rate vs. bandwidth, spread spectrum communications, and modulation methods.

E8C Test Notes

- Morse code is a digital code that consists of elements having unequal length.
- Baudot digital code and ASCII differ in that Baudot uses five data bits per character, while ASCII uses seven or eight; Baudot uses two characters as shift codes, and ASCII has no shift code.
- Using the ASCII code for data communications makes it possible to transmit both upper- and lowercase text.
- Minimizing the bandwidth requirements of a PSK31 can be accomplished with use of sinusoidal data pulses.
- The bandwidth needed for a 13 WPM international Morse code transmission is approximately 52 Hz.
- The bandwidth needed for a 170 Hz shift, 300-baud ASCII transmission is 0.5 kHz.
- The bandwidth needed for a 4800 Hz frequency shift, 9600-baud ASCII FM transmission is 15.36 kHz.
- Spread spectrum communication is a wide-bandwidth communications system in which the transmitted carrier frequency varies according to some predetermined sequence.
- Spread spectrum causes a digital signal to appear as wide-band noise to a conventional receiver.
- Frequency hopping alters the center frequency of a conventional carrier many times per second in accordance with a pseudo-random list of channels.
- Direct sequencing uses a high-speed binary bit stream to shift the phase of an RF carrier.

- Some types of errors can be detected when including a parity bit with an ASCII character stream.
- An advantage of using JT-65 coding is the ability to decode signals that have a very low signal-to-noise ratio.

Group E8D – Waves, Measurements, and RF Grounding

Questions in group E8D cover peak-to-peak values, polarization, and RF grounding

E8D Test Notes

- Peak-to-peak voltage is easiest to measure when viewing a pure sine wave signal on an analog oscilloscope.
- The ratio between the peak-to-peak voltage and peak voltage amplitudes in a symmetrical waveform is 2:1.
- Peak voltage is valuable in evaluating the signal-handling capability of a Class A amplifier.
- The PEP output of a transmitter that develops a peak voltage of 30 volts into a 50-ohm load is 9 watts.
- If an RMS-reading AC voltmeter reads 65 volts on a sinusoidal waveform, the peak-to-peak voltage is 184 volts.
- Using a peak-reading wattmeter to monitor the output of an SSB phone transmitter gives a more accurate display of the PEP output when modulation is present.
- An electromagnetic wave consists of an electric field and a magnetic field oscillating at right angles to each other.
- Changing electric and magnetic fields propagate the energy of electromagnetic waves traveling in free space.
- Circularly polarized electromagnetic waves have a rotating electric field.
- A peak-reading wattmeter should be used to monitor the output signal of a voice-modulated single-sideband transmitter to ensure you do not exceed the maximum allowable power.
- The average power dissipated by a 50-ohm resistive load during one complete RF cycle with a peak voltage of 35 volts is 12.2 watts.

- If an RMS-reading voltmeter reads 34 volts, peak voltage of the measured sinusoidal waveform is 48 volts.
- Peak voltage at a standard U.S. household electrical outlet is typically 170 volts.
- Peak-to-peak voltage at a standard U.S. household electrical outlet is typically 340 volts.
- RMS voltage at a standard U.S. household electrical power outlet is typically 120V AC.
- RMS value of a 340-volt peak-to-peak pure sine wave is 120V AC.

5.9 SUBELEMENT E9

Subelement E9 contains eight targeted question groups with a total of 109 potential test questions. Eight of the 50 [Element 4] test questions will be drawn from subelement E9.

Group E9A – Isotropic and Gain Antennas

Questions in group E9A cover the following topics: definitions, used as a standard for comparison, radiation pattern, basic antenna parameters, radiation resistance and reactance, gain, beamwidth, and efficiency.

E9A Test Notes

- An isotropic antenna is a theoretical antenna used as a reference for antenna gain.
- A 1/2-wavelength dipole in free space has 2.15 dB of gain.
- An isotropic antenna has no gain in any direction.
- The feed-point impedance of an antenna is needed to match impedances in order to minimize standing-wave ratio on the transmission line.
- Antenna height, conductor length/diameter ratio, and location of nearby conductive objects may affect the feed-point impedance of an antenna.
- The total resistance of an antenna system includes radiation resistance plus ohmic resistance.

- A folded dipole antenna is a dipole constructed from one wavelength of wire forming a very thin loop.
- Antenna gain is the ratio relating the radiated signal strength of an antenna in the direction of maximum radiation to that of a reference antenna.
- Antenna bandwidth is the frequency range over which an antenna satisfies a performance requirement.
- The formula for calculating antenna efficiency is (radiation resistance / total resistance) x 100%.
- Installing a good radial system improves the efficiency of a ground-mounted 1/4-wave vertical antenna.
- Soil conductivity determines ground losses for a ground-mounted vertical antenna operating in the 3 – 30 MHz range.
- An antenna has 3.85 dB of gain compared to a 1/2-wavelength dipole when it has 6 dB gain over an isotropic antenna.
- An antenna has 9.85 dB of gain compared to a 1/2-wavelength dipole when it has 12 dB gain over an isotropic antenna.
- The radiation resistance of an antenna is the value of a resistance that would dissipate the same amount of power as radiated from an antenna.

Group E9B – Antenna Patterns

Questions in group E9B cover the following topics: E and H plane patterns, gain as a function of pattern, antenna design, plus Yagi antennas. Some questions from this group are based on the figure shown below.

Figure E9-1

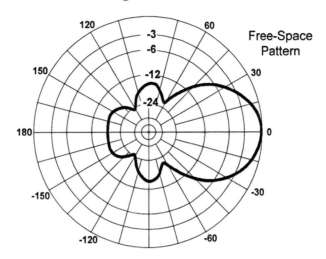

E9B Test Notes

- In the antenna radiation pattern shown in Figure E9-1, the 3-dB beamwidth is 50 degrees.
- In the antenna radiation pattern shown in Figure E9-1, the front-to-back ratio is 18 dB.
- In the antenna radiation pattern shown in Figure E9-1, the front-to-side ratio is 14 dB.
- When a directional antenna is operated at different frequencies within the band for which it was designed, the gain may change depending on frequency.
- The front-to-back ratio decreases if a Yagi antenna is designed solely for maximum forward gain.
- Gain increases if the boom of a Yagi antenna is lengthened and the elements are properly retuned.
- Assuming each is driven by the same amount of power, the total amount of radiation emitted by a directional gain antenna is the same as the total amount of radiation emitted from an isotropic antenna.
- The approximate beamwidth in a given plane of a directional antenna can be determined by noting the two points where the signal strength of the antenna is 3 dB less than maximum and computing the angular difference.

- Method of moments is a computer program technique commonly used for antenna modeling.
- In Method of moments analysis, a wire is modeled as a series of segments with a uniform value of current for each.
- The computed feed-point impedance may be incorrect if the number of wire segments in an antenna model is decreased below the guideline of 10 segments per 1/2 wavelength.
- The far-field of an antenna is the region where the shape of the antenna pattern is independent of distance.
- When applied to antenna modeling programs, NEC stands for Numerical Electromagnetics Code.
- SWR vs. frequency charts, polar plots of the far-field elevation and azimuth patterns, plus antenna gain can all be obtained through the use of antenna modeling programs.

Group E9C – Wire and Phased Vertical Antennas

Questions in group E9C cover the following topics: Beverage antennas, terminated and resonant rhombic antennas, elevation above real ground, ground effects as related to polarization, and take-off angles. Some questions from this group are based on the figure shown below.

Figure E9-2

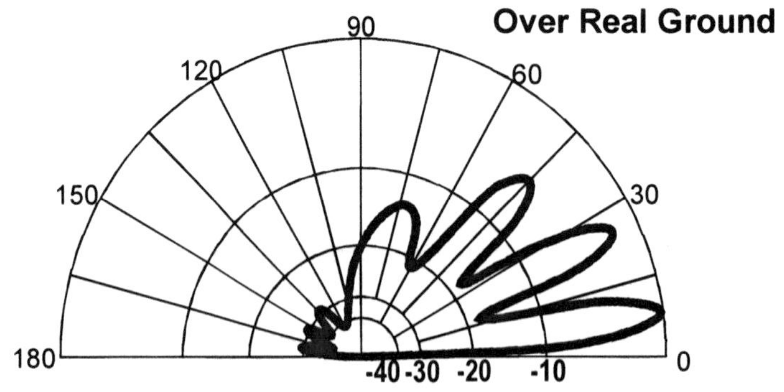

E9C Test Notes

- The radiation pattern of two 1/4-wavelength vertical antennas spaced 1/2 wavelength apart and fed 180 degrees out of phase is a figure eight oriented along the axis of the array.

- The radiation pattern of two 1/4-wavelength vertical antennas spaced 1/4 wavelength apart and fed 90 degrees out of phase is a cardioid.

- The radiation pattern of two 1/4-wavelength vertical antennas spaced 1/2 wavelength apart and fed in phase is a figure eight broadside to the axis of the array.

- A basic unterminated rhombic antenna is bidirectional with four sides, each side one or more wavelengths long, and open at the end opposite the transmission line connection.

- A terminated rhombic antenna for the HF bands requires a large physical area and four separate supports.

- A terminating resistor on a rhombic antenna changes the radiation pattern from bidirectional to unidirectional.

- Figure E9-2 shows an elevation antenna pattern over real ground.

- The elevation angle of peak response shown in Figure E9-2 is 7.5 degrees.

- The front-to-back ratio shown in Figure E9-2 is 28 dB.

- Four elevation lobes appear in the forward direction of Figure E9-2.

- The low-angle radiation increases when a vertically polarized antenna is mounted over seawater instead of rocky ground.

- A Beverage antenna should be one or more wavelengths long to achieve good performance at the desired frequency.

- Placing a vertical antenna over an imperfect ground reduces low-angle radiation.

Group E9D – Directional Antennas

Questions in group E9D cover the following topics: gain, satellite antennas, antenna beamwidth, antenna losses, SWR bandwidth, antenna efficiency, shortened and mobile antennas, and grounding.

E9D Test Notes

- When the operating frequency is doubled, the gain of an ideal parabolic dish antenna increases by 6 dB.
- To produce circular polarization, arrange two linearly polarized Yagi antennas perpendicular to each other with the driven elements at the same point on the boom and fed 90 degrees out of phase.
- The beamwidth of an antenna decreases as gain is increased.
- It desirable for a ground-mounted satellite communications antenna system to be able to move in both azimuth and elevation in order to track the satellite as it orbits the Earth.
- A high-Q loading coil should be placed near the center of the vertical radiator to minimize losses in a shortened vertical antenna.
- An HF mobile antenna loading coil should have a high ratio of reactance to resistance to minimize losses.
- A multiband trapped antenna might radiate harmonics.
- The bandwidth of an antenna decreases as it is shortened through the use of loading coils.
- An advantage of using top loading in a shortened HF vertical antenna is improved radiation efficiency.
- The feed-point impedance at the center of a two-wire folded dipole antenna is approximately 300 ohms.
- A loading coil can be used with an HF mobile antenna to cancel capacitive reactance.
- A trapped antenna may be used for multiband operation.
- As the frequency of operation is lowered, radiation resistance decreases and capacitive reactance increases at the base of a fixed-length HF mobile antenna.
- A wide, flat copper strap is the type of conductor for minimizing losses in a station's RF ground system.

- An electrically-short connection to 3 or 4 interconnected ground rods driven into the Earth will provide the best RF ground for an amateur radio station.

Group E9E – Matching

Questions in group E9E cover matching antennas to feed lines and power dividers.

E9E Test Notes

- The delta matching system matches a high-impedance transmission line to a lower impedance antenna by connecting the line to the driven element in two places spaced a fraction of a wavelength each side of element center.
- The gamma match system matches an unbalanced feed line to an antenna by feeding the driven element both at the center of the element and at a fraction of a wavelength to one side of center.
- The stub match system uses a section of transmission line connected in parallel with the feed line at or near the feed point.
- The series capacitor in a gamma-type antenna matching network cancels the inductive reactance of the matching network.
- The driven element reactance must be capacitive in a three-element Yagi to use a hairpin matching system.
- An L network is the equivalent lumped-constant network for a hairpin matching system on a three-element Yagi.
- Reflection coefficient describes the interactions at the load end of a mismatched transmission line.
- An SWR greater than 1:1 is characteristic of a mismatched transmission line.
- Gamma match is an effective method of connecting a 50-ohm coaxial cable feed line to a grounded tower so it can be used as a vertical antenna.
- Inserting a 1/4-wavelength piece of 75-ohm coaxial cable transmission line in series between the antenna terminals and the 50-ohm feed cable is an effective way to match an antenna with a 100-ohm feed-point impedance to a 50-ohm coaxial cable feed line.

- The universal stub matching technique is an effective way of matching a feed line to a VHF or UHF antenna when the impedances of the antenna and feed line are unknown.
- Phasing line, when used with an antenna having multiple driven elements, ensures each driven element operates in concert with the others to create the desired antenna pattern.
- A Wilkinson divider divides power equally among multiple loads while preventing changes in one load from disturbing power flow to the others.

Group E9F – Transmission Lines

Questions in group E9F cover the following topics: characteristics of open and shorted feed lines, 1/8 wavelength, 1/4 wavelength, 1/2 wavelength, feed lines, coax versus open-wire, velocity factor, electrical length, transformation characteristics of line terminated in impedance not equal to characteristic impedance.

E9F Test Notes

- The velocity factor of a transmission line is the velocity of the wave in the transmission line divided by the velocity of light in a vacuum.
- The dielectric materials used in a transmission line determine the velocity factor.
- The physical length of a coaxial cable transmission line is shorter than its electrical length because electrical signals move more slowly in a coaxial cable than in air.
- The typical velocity factor for a coaxial cable with solid polyethylene dielectric is 0.66.
- The physical length of a solid polyethylene dielectric coaxial transmission line that is electrically 1/4 wavelength long at 14.1 MHz is approximately 3.5 meters.
- The physical length of an air-insulated, parallel conductor transmission line that is electrically 1/2 wavelength long at 14.10 MHz is approximately 10 meters.
- Ladder line has lower loss compared to small-diameter coaxial cable.

- Velocity factor is the ratio of the actual speed at which a signal travels through a transmission line to the speed of light in a vacuum.
- The physical length of a solid polyethylene dielectric coaxial transmission line that is electrically 1/4 wavelength long at 7.2 MHz is approximately 6.9 meters.
- A 1/8-wavelength transmission line presents an inductive reactance to a generator when the line is shorted at the far end.
- A 1/8-wavelength transmission line presents a capacitive reactance to a generator when the line is open at the far end.
- A 1/4-wavelength transmission line presents a very low impedance to a generator when the line is open at the far end.
- A 1/4-wavelength transmission line presents a very high impedance to a generator when the line is shorted at the far end.
- A 1/2-wavelength transmission line presents a very low impedance to a generator when the line is shorted at the far end.
- A 1/2-wavelength transmission line presents a very high impedance to a generator when the line is open at the far end.
- Assuming all other parameters are the same, the differences between foam-dielectric coaxial cable and solid-dielectric cable are reduced safe operating voltage limits, reduced losses per unit of length, and higher velocity factor.

Group E9G – Antennas and Transmission Lines

Questions in group E9G cover the Smith chart. Some questions from this group are based on the figure shown below.

Figure E9-3

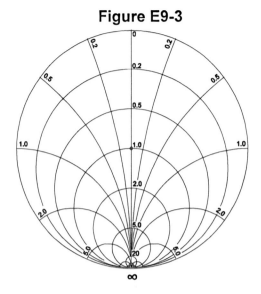

E9G Test Notes

- Impedance along transmission lines can be calculated using a Smith chart.
- Resistance circles and reactance arcs are used for coordinates in a Smith chart.
- Impedance and SWR values in transmission lines are often determined using a Smith chart.
- Resistance and reactance are the two families of circles and arcs that make up a Smith chart.
- A Smith chart is shown in Figure E9-3.
- On the Smith chart shown in Figure E9-3, the large outer circle on which the reactance arcs terminate is called the reactance axis.
- On the Smith chart shown in Figure E9-3, the straight line is called the resistance axis.
- With a Smith chart, the process of normalization is reassigning impedance values with regard to the prime center.
- Standing-wave ratio circles are often added to a Smith chart during the process of solving problems.
- The arcs on a Smith chart represent points with constant reactance.

- The wavelength scales on a Smith chart are calibrated in fractions of transmission line electrical wavelength.

Group E9H – Antennas and Transmission Lines

Questions in group E9H cover the following topics: effective radiated power, system gains and losses, plus radio direction-finding antennas.

E9H Test Notes

- The effective radiated power relative to a dipole of a repeater station with 150 watts transmitter power output, 2-dB feed line loss, 2.2-dB duplexer loss and 7-dBd antenna gain is 286 watts.
- The effective radiated power relative to a dipole of a repeater station with 200 watts transmitter power output, 4-dB feed line loss, 3.2-dB duplexer loss, 0.8-dB circulator loss and 10-dBd antenna gain is 317 watts.
- The effective isotropic radiated power of a repeater station with 200 watts transmitter power output, 2-dB feed line loss, 2.8-dB duplexer loss, 1.2-dB circulator loss and 7-dBi antenna gain is 252 watts.
- Effective radiated power describes station output, including the transmitter, antenna, and everything in between, when considering transmitter power and system gains and losses.
- The main drawback of a wire-loop antenna for direction finding is that it has a bidirectional pattern.
- In the triangulation method of direction finding, antenna headings from several different receiving locations are used to locate the signal source.
- An RF attenuator on a receiver being used for direction finding prevents receiver overload, which could make it difficult to determine peaks or nulls.
- A sense antenna modifies the pattern of a DF antenna array to provide a null in one direction.
- Receiving loop antennas are constructed with one or more turns of wire wound in the shape of a large open coil.

- The output voltage of a multi-turn receiving loop antenna can be increased by increasing either the number of wire turns in the loop, the area of the loop structure, or both.
- The very sharp single null of a cardioid-pattern antenna is useful for direction finding.
- A shielded loop antenna is useful for direction finding because it is electrostatically balanced against ground, giving better nulls.

5.10 SUBELEMENT E0

Subelement E0 contains 1 targeted question group with a total of 11 potential test questions. One of the 50 [Element 4] test questions will be drawn from subelement E0.

Group E0A – Safety

Questions in group E0A cover amateur radio safety practices, RF radiation hazards, and hazardous materials.

E0A Test Notes

- Radioactive materials emit ionizing radiation, while RF signals have less energy and can only cause heating.
- When evaluating RF exposure levels from your station at a neighbor's home, make sure signals from your station are less than the uncontrolled MPE limits.
- To estimate whether the RF fields produced by an amateur radio station are within permissible MPE limits, use an antenna modeling program to calculate field strength at accessible locations.
- When evaluating a site with multiple transmitters operating at the same time, each transmitter that produces 5% or more of its MPE exposure limit at accessible locations is responsible for mitigating over-exposure situations.

- The high-gain antennas commonly used in the amateur radio microwave bands can result in high exposure levels.
- There are separate electric (E) and magnetic (H) field MPE limits because the body reacts to electromagnetic radiation from both the E and H fields, ground reflections and scattering make the field impedance vary with location, plus E field and H field radiation intensity peaks can occur at different locations.
- Dangerous levels of carbon monoxide from an emergency generator can be detected only with a carbon monoxide detector.
- SAR measures the rate at which RF energy is absorbed by the body.
- Beryllium oxide is extremely toxic when broken or crushed and the particles accidentally inhaled.
- Polychlorinated biphenyl is used in some electronic components, such as high-voltage capacitors and transformers, and is considered toxic.
- Using high-power UHF or microwave transmitters can result in localized heating of the body from RF exposure in excess of the MPE limits.

5.11 STUDY TIPS

Review the test notes for each subelement carefully. Once you are comfortable with the majority of information presented, take the practice exam in Appendix 3 to check your progress. Repeat the process as needed.

You should also try a random sampling of the Element 4 question pool for more practice. The resource directory in Appendix 4 will show you not only where to download the entire question pool online but also where to find detailed explanations for all material covered on the licensing exam. Everything you need to pass your Amateur Radio (Ham) License Examination is in this manual – all you have to do is use the information, study with diligence, and practice!

APPENDIX 1
Element 2 Practice Exam

1. What is required when you operate at 1296.1 MHz and discover that you are causing interference with a radiolocation terminal stationed outside U.S. borders?
A. Get rid of the interference or quit operating
B. Ignore all interference because the frequency is allocated for amateur radio
C. Call out to the terminal operator and request RF tuning
D. Use Morse code because it won't interfere

2. Which amateur band allows Technician Class licensees to operate at a frequency of 434.500 MHz?
A. 23 centimeter
B. 33 centimeter
C. 70 centimeter
D. None of the answers are correct

3. What kinds of worldwide communications are allowed by licensed amateur stations?
A. Types associated with the amateur service and comments of a personal character
B. Types associated with international business and comments of a personal nature
C. Types associated FCC-sponsored contests only
D. All communications allowed by global broadcast stations

4. When is an amateur station permitted to encrypt a message?
A. Only when participating in international competition
B. Only during mobile operation
C. Only for control commands to space stations and radio controlled vehicles
D. Only at frequencies above 1.1 GHz

5. Which best demonstrates automatic control?
A. Repeater functions
B. Web-based station control
C. Sending blog updates with a cell phone
D. Sending Morse code with a cell phone

6. When must a station call sign be followed by the /AG indicator?
A. During mobile operation
B. During repeater operation
C. While operating outside the license region of origin
D. While using new privileges earned before the FCC database shows a license upgrade

7. How often is station identification required during a transmit test?
A. Every 90 seconds
B. Every 2 minutes
C. Every 5 minutes
D. Every 10 minutes

8. What type of communication uses the same frequency to transmit and receive?
A. Duplex
B. Diplex
C. Simplex
D. Multiplex

9. What is the recommended method for contacting an NCS to report an emergency?
A. Announce "SOS" or "Mayday" three times before your call sign
B. Transmit a 1 KHz emergency alert signal before your call sign
C. Announce "Emergency" or "Priority" before your call sign
D. Press your PTT button three times before announcing your call sign

10. In what part of the atmosphere are radio waves able to propagate worldwide?
A. Electrosphere
B. Ionosphere
C. Stratosphere
D. Troposphere

11. Which are the primary building blocks for radio waves?
A. Power and voltage
B. Resistance and current
C. Electric and magnetic fields
D. Ionized and de-ionized radioactivity

12. What is most likely to carry VHF radio signals over unusually long distances?
A. Transient lightning
B. Increased solar wind
C. High-altitude jet stream
D. Sporadic E propagation

13. Which is the best location to check SWR?
A. Between a station's transmitter and antenna
B. At a station's main grounding bus
C. In series with a station's input
D. In parallel with a station's input

14. The acronym "RIT" refers to which function?
A. Remote Input Testing
B. Receiver Incremental Tuning
C. Random Inverter Tone
D. Rectified In-line Transmission

15. Electron flow is caused by which electromotive force?
A. Voltage
B. Resistance
C. Inductance
D. Capacitance

16. 2.5 amperes equals how many milliamperes?
A. 25 milliamperes
B. 250 milliamperes
C. 2500 milliamperes
D. 25000 milliamperes

17. Which electrical property uses magnetic fields for energy storage?
A. Resistance
B. Capacitance
C. Conductance
D. Inductance

18. How much current would you expect to find in a 100-ohm circuit with a 110-volt supply?
A. 0.11 amps
B. 1.1 amps
C. 11 amps
D. 110 amps

19. Which component stores energy in an electrostatic field?
A. Diode
B. Resistor
C. Capacitor
D. Inductor

20. Which component can be used for a signal amplifier?
A. Transistor
B. Thermistor
C. Shottky diode
D. Varactor diode

21. What are represented by the symbols on an electrical schematic diagram?
A. Circuit components
B. Digital tolerances
C. Logic states
D. Signal flow

22. What is a typical use for tuned circuits?
A. Filtering
B. Resistive loads
C. Voltage regulation
D. High-fidelity reproduction

23. Which process best describes combining audio with an RF carrier?
A. High-pass filtering
B. Demultiplexing
C. Modulation
D. Oscillation

24. Which of the following is NOT used for eliminating RF interference?
A. Low-pass filters
B. High-pass filters
C. Ferrite chokes
D. Steering diodes

25. What type of procedure would you perform to check if a station load is matched to a transmission line?
A. Standing-wave ratio test
B. Station ground test
C. Transmitter alignment check
D. Receiver alignment check

26. What are telltale signs of a cold solder joint?
A. Chemical residue and discoloration
B. Smooth, shiny surfaces
C. Dull, pitted surfaces
D. Residual smell of burnt components

27. Which is the most popular modulation type used for packet radio in the VHF range?
A. Amplitude modulation
B. Frequency modulation
C. Single-sideband
D. Spread spectrum

28. What causes Doppler shift during satellite communications?
A. The satellite transmits and receives on different frequency bands
B. The satellite changes orbit
C. The digital communications mode changes
D. The signal frequency changes

29. Where might you find a listing of active VoIP nodes?
A. A repeater directory
B. A repeater frequency coordinator
C. An emergency coordinator
D. The FCC Web site

30. Which is a form of digital communications?
A. Multiple frequency-shift keying
B. Phase-shift keying
C. Packet radio
D. All of the above

31. What is the approximate wavelength in meters for a frequency of 100 MHz?
A. 0.3 meters
B. 3 meters
C. 30 meters
D. 300 meters

32. Why is low SWR important?
A. It extends the life of an antenna
B. It reduces loss
C. It reduces interference
D. All of the above

33. What is a good precaution when installing lightning protection for feed lines?
A. Install a disconnect switch in series with the ground line for each protection device
B. Install a bypass switch in parallel with each protection device
C. Ground each protection device separately to equipment ground
D. Ground all protection devices to a common external ground point

34. Which are used to lift tower sections and antennas?
A. Ground straps
B. Guy wires
C. Gin poles
D. Guide pins

35. How is RF radiation different from ionizing radiation?
A. RF radiation does not cause genetic damage
B. RF radiation causes genetic damage
C. RF radiation is not dangerous and poses no health risks
D. None of the above

ANSWER KEY
1. **A** | FCC wants you to eliminate harmful interference or cease operating.
2. **C** | Licensed amateurs may operate at 430 – 440 MHz of the 70 cm band in all ITU Regions.
3. **A** | FCC permits communications incidental to the purposes of the amateur service and remarks of a personal character.
4. **C** | Otherwise transmitting encrypted messages is not permitted.
5. **A** | Repeater functions best show automatic control; the remaining operations require other manual input.
6. **D** | /KT, /AG or /AE is used when operating under new privileges while a license upgrade is pending update in the FCC license database.
7. **D** | Station identification is needed at least every 10 minutes and when the test is done.
8. **C** | Simplex communication indicates use of the same frequency for transmit and receive.
9. **C** | Get the attention of a net control station by announcing "Emergency" or "Priority" before your call sign.
10. **B** | Radio signals can propagate globally in the ionosphere.
11. **C** | Radio waves are comprised of electric and magnetic fields.
12. **D** | The atmosphere's E layer characteristics sometimes allow long-distance VHF communications.
13. **A** | Check standing-wave ratio at the feed line between a transmitter and antenna.
14. **B** | Offset a receiver frequency from the transmitter with receiver incremental tuning.
15. **A** | Voltage is the electromotive force, or EMF, that causes the flow of electrons.
16. **C** | 2500 milliamperes is the same as 2.5 amperes.
17. **D** | Inductance opposes change in electric current and uses electromagnetism to induce voltage in neighboring conductors.
18. **B** | Per Ohm's Law, 110 volts divided by 100 ohms of resistance equals 1.1 amps of current.
19. **D** | Capacitors use plates and a non-conductive dielectric to store electrostatic energy.
20. **A** | Transistors are used for both switching and amplification.
21. **A** | Symbols on a schematic diagram represent electrical components in the circuit.
22. **A** | Normally comprised of inductors and capacitors, tuned circuits are band-pass filters.
23. **C** | Modulation is the process of altering characteristics of a carrier signal with other information.
24. **D** | Filters and chokes are commonly used to eliminate interference from unwanted frequencies.
25. **A** | An SWR test indicates how well a load is matched to a transmission line by measuring the ratio between forward and reflected power.
26. **C** | Cold solder joints often leave a dull, pitted appearance.

27. **B** | VHF packet radio normally uses frequency modulation for digital data transmission.
28. **D** | Relative motion between an Earth station and satellite can cause observable changes in   signal frequency.
29. **A** | Repeater directories often contain a list of active nodes utilizing VoIP.
30. **D** | All are forms of digital communications.
31. **B** | The wavelength of 100 MHz is approximately 3 meters.
32. **B** | Low SWR allows efficient transmission and reduces loss associated with high reflected   power.
33. **D** | Grounding all lightning protection devices to a common external ground point provides   the best station protection.
34. **C** | Gin poles are used to lift tower sections and antennas.
35. **A** | RF radiation does not produce the kinetic energy required to cause genetic damage.

APPENDIX 2
Element 3 Practice Exam

1. General Class licensees are granted all amateur privileges on which bands?
A. 12, 17, 20, and 60 meter
B. 10, 40, 80, and 160 meter
C. 10, 12, 17, 30, 60, and 160 meter
D. 10, 12, 15, 17, 30, and 160 meter

2. When erected away from public airports, what is the maximum allowable height for antenna structures without notifying the FAA?
A. 300 feet
B. 200 feet
C. 100 feet
D. 50 feet

3. What is the transmitter PEP limit when operating in the 28 MHz band?
A. 500 watts
B. 1000 watts
C. 1500 watts
D. 2000 watts

4. Who accredits Volunteer Examiners?
A. FCC Officers
B. Wireless Telecommunications Bureau Reps
C. Volunteer Examiner Coordinators
D. Universal Licensing System Managers

5. When must a licensee take specific actions to avoid harmful interference?
A. When operating within a mile of FCC monitoring stations
B. When the amateur service is considered secondary
C. When transmitting spread spectrum emissions
D. All of the above

6. What voice mode is most common on the high-frequency amateur bands?
A. FM
B. DSB-SC
C. SSB
D. PM

7. What action should you take when a station in distress breaks in on your communications?
A. Finish your communication then maintain radio silence
B. Acknowledge the distressed station and decide what type of help may be required
C. Continue your communications on another frequency
D. Stop transmitting right away

8. What is QSK?
A. Fully automated Morse code
B. Fully automated voice communications
C. Full duplex operation
D. Full break-in radiotelegraphy

9. Where is a directional antenna aimed for long-path contact?
A. 90 degrees from the antenna's short-path bearing
B. 180 degrees from the antenna's short-path bearing
C. 28.5 degrees above the north horizon
D. 28.5 degrees above the south horizon

10. Where will you find routing and handling information?
A. In a data packet directory
B. In a data packet preamble
C. In a data packet header
D. In a data packet footer

11. What is a potential result of increased geomagnetic activity?
A. An auroral zone that reflects VHF signals
B. Increased high frequency signal strength at the polar regions
C. Better high frequency long-path propagation
D. Reduction of extended delay echo

12. Which is a common occurrence when operating below the lowest usable frequency?
A. Radio signals get reflected by the ionosphere
B. Radio signals are trapped in the ionosphere
C. Radio signals are absorbed by the ionosphere
D. Radio signals are passed through the ionosphere

13. When do layers of the ionosphere reach maximal height?
A. When the sun is overhead
B. When the sun is on the reverse side of the planet
C. At dawn
D. At dusk

14. Which of the following are used for impedance matching?
A. Balanced modulators
B. SWR bridges
C. Antenna couplers
D. Q multipliers

15. Which of the following have horizontal and vertical channels?
A. Ohmmeters
B. Signal generators
C. Ammeters
D. Oscilloscopes

16. What would you use a digital signal processor for in an amateur station?
A. Improving station grounding
B. Removing noise from receive signals
C. Increasing antenna gain
D. Increasing antenna bandwidth

17. An S meter is part of which amateur station subsystem?
A. Transmitter
B. Receiver
C. SWR bridge
D. Conductance bridge

18. Which is more likely to happen when using short mobile antennas instead of full-sized antennas?
A. Transmit distortion
B. Signal polarization
C. Limited bandwidth
D. Increased harmonics

19. What is the result of equal impedance matching between power source and electrical load?
A. Decreased current through the circuit
B. A short in the electrical load
C. Minimum power transfer from source to load
D. Maximum power transfer from source to load

20. What is the RMS value of a 27-volt (peak) sine wave?
A. 13.5 volts
B. 19.1 volts
C. 37.8 volts
D. 54.0 volts

21. What is the total resistance of four 220-ohm resistors in parallel?
A. 55 ohms
B. 110 ohms
C. 440 ohms
D. 880 ohms

22. What is a benefit of using ceramic capacitors instead of other types?
A. Tighter tolerance
B. Quicker charge time
C. Quicker discharge time
D. Lower cost

23. What is the approximate junction threshold voltage of conventional silicon diodes?
A. 0.3 volts
B. 0.5 volts
C. 0.7 volts
D. 0.9 volts

24. What is non-volatile memory?
A. Stored data is not affected by radio frequencies
B. Stored data is not affected by extreme heat
C. Stored data is not lost when power is removed
D. Stored data is not permanent and can be overwritten

25. What are the primary components found in most filter circuits?
A. Transistors and diodes
B. Capacitors and inductors
C. Resistors and transformers
D. Registers and logic gates

26. Which of the following are most efficient?
A. Class A amplifiers
B. Class B amplifiers
C. Class AB amplifiers
D. Class C amplifiers

27. Which combination of subsystems make up a complete basic superheterodyne receiver?
A. RF amplifier, detector, audio amplifier
B. RF amplifier, mixer, IF discriminator
C. HF oscillator, mixer, detector
D. HF oscillator, pre-scaler, audio amplifier

28. Which is a consequence of over-modulating a signal?
A. Unstable frequency lock
B. Not enough audio
C. Not enough bandwidth
D. Too much bandwidth

29. What is the result of matching bandwidth between operating mode and receiver?
A. Best possible impedance matching
B. Reduced frequency drift
C. Decreased power consumption
D. Best possible signal-to-noise ratio

30. Which of the following is a typical expression of RF feed line loss?
A. ohms per 100 feet
B. ohms per 1000 feet
C. dB per 100 feet
D. dB per 1000 feet

31. Which of the following is a potential drawback from using directly fed random-wire antennas?
A. Excessive size requirements
B. Harmful RF hazards
C. Lack of horizontal radiation
D. Less effective at higher frequencies

32. Which best describes an antenna's main lobe?
A. The maximal vertical angle of radiation
B. The point of maximal antenna element current
C. The point of maximal standing-wave voltage
D. The direction of maximal radiated field strength

33. Which of the following best describes the main reason to use an antenna trap?
A. Allow multi-band operation
B. Cut away spurious frequencies
C. Provide impedance matching
D. Band-stop filtering

34. Which is the most accurate type of equipment for measuring an RF field?
A. An S meter
B. A field-strength meter with calibrated antenna
C. A betascope with a calibrated dummy load
D. An oscilloscope

35. When is a lead-acid battery most likely to leak explosive hydrogen gas?
A. During extended periods in storage
B. While discharging
C. While charging
D. If stored on an uneven surface

ANSWER KEY

1. **C** | General Class licensees are granted all amateur privileges on these frequency bands.
2. **B** | FAA notification and FCC registration are required when antenna structures exceed 200 feet in height.
3. **C** | Transmitter power is limited to 1500 watts PEP in the 28 MHz band.
4. **C** | Volunteer Examiners are accredited by Volunteer Examiner Coordinators.
5. **D** | Licensees must take specific actions to avoid harmful interference to others under all listed circumstances.
6. **C** | Single-sideband is the most common voice communications mode on the high-frequency amateur bands.
7. **B** | Acknowledge any distressed station and determine what assistance may be needed.
8. **D** | QSK is full break-in radiotelegraphy.
9. **B** | For long-path contact, aim a directional antenna 180 degrees from its short-path bearing.
10. **C** | A data packet header contains routing and handling information.
11. **A** | High geomagnetic activity can result in an aurora that reflects VHF signals.
12. **C** | Operating below the lowest usable frequency normally results in radio signals getting totally absorbed by the ionosphere.
13. **A** | Ionospheric layers reach maximal height when the sun is overhead.
14. **C** | Antenna couplers are useful for impedance matching when coupling to something other than 50 ohms.
15. **D** | Oscilloscopes utilize both horizontal and vertical channel amplifiers.
16. **B** | Digital signal processors can help remove noise from receive signals in amateur stations.
17. **B** | Signal strength meters are indicators found on most communication receivers.
18. **C** | Full-sized mobile antennas are not as limited in bandwidth as their shortened counterparts.
19. **D** | Equal impedance matching between source and load always results in maximum transfer of power.
20. **B** | 27 volts peak multiplied by 70.7 percent equals 19.1 volts RMS.
21. **A** | The total equivalent resistance of four 220-ohm resistors in parallel is 55 ohms.
22. **D** | Ceramic generally cost less than other types of capacitors.
23. **C** | The junction threshold for most conventional silicone diodes is approximately 0.7 volts of forward bias.
24. **C** | Non-volatile memory is NOT lost when power is removed and data remains in storage.
25. **B** | Capacitors and inductors are typically the main components used for filtering circuits.
26. **D** | With up to 90% efficiency, Class C amplifiers are the most efficient of all listed circuits.

27. **C** | A basic superheterodyne receiver is comprised of an HF oscillator, a mixer, and a detector.
28. **D** | Over-modulation often results in excessive use of bandwidth.
29. **D** | Matching bandwidth between operating mode and receiver often yields the best possible signal-to-noise ratio.
30. **C** | RF feed line loss is normally expressed in dB per 100 feet.
31. **B** | Coming in contact with metal objects in the station can result in RF burns.
32. **D** | The direction of maximal radiated field strength is a directive antenna's main lobe.
33. **A** | Antenna traps are mainly used to allow multi-band operation.
34. **B** | A calibrated field-strength meter and antenna provide the most accuracy for measuring RF fields.
35. **C** | Lead-acid batteries are most likely to leak explosive hydrogen gas while charging.

APPENDIX 3
Element 4 Practice Exam

1. Which amateur band has specific transmit channels instead of a frequency range?
A. 12 meter
B. 17 meter
C. 30 meter
D. 60 meter

2. How do you obtain authorization to place an amateur station within a wildlife preserve?
A. Submit a proposal to the National Park Service
B. File a letter of intent with the National Audubon Society
C. Submit an environmental assessment to the FCC
D. Submit an FSD-15 form to the Department of the Interior

3. Which of the following are allowed automatic retransmission of radio signals from other stations?
A. Beacon, repeater and space stations
B. Auxiliary, repeater and space stations
C. Earth and repeater stations only
D. Auxiliary, beacon and space stations

4. Which amateur license class authorizes someone to be the control operator of a space station?
A. Technician
B. General
C. Amateur Extra
D. All of the above

5. What is the minimum passing test score needed to receive an amateur operator license?
A. 72%
B. 74%
C. 76%
D. 78%

6. Where is the U.S. National Radio Quiet Zone located?
A. In New Mexico near the White Sands Test Area
B. In Washington, D.C., near the Naval Research Laboratory
C. In West Virginia near the National Radio Astronomy Observatory
D. In Florida near Cape Canaveral

7. What type of information could you use to predict the location of a satellite?
A. Doppler data
B. Base PPI Reflectivity
C. Composite Reflectivity
D. Keplerian elements

8. What is the transmitted frame rate for NTSC television?
A. 30 fps
B. 60 fps
C. 90 fps
D. 120 fps

9. Why are spread spectrum techniques less prone to interference?
A. Spread spectrum receivers suppress radio signals without the correct algorithm
B. Spread spectrum transmitters use higher power
C. Spread spectrum receivers are equipped with digital blanker circuits
D. Spread spectrum receivers force transmitters to switch frequencies upon detecting interference

10. What is baud rate?
A. The number of data symbols per second
B. The number of characters per second
C. The number of characters per minute
D. The number of words per minute

11. What is the most popular data transmission method below 30 MHz?
A. FM
B. FSK
C. Pulse modulated
D. Spread spectrum

12. In which state of the moon are you likely to experience the least amount of path loss during EME communications?
A. During a new moon phase
B. During a full moon phase
C. At perigee
D. At apogee

13. Which could cause an echo in receive signals?
A. D-layer absorption
B. Meteor bursts
C. Transmitting above the maximum usable frequency
D. Receiving a radio signal from multiple paths

14. What happens to the maximum distance of ground-wave propagated signal when its frequency is increased?
A. Remains the same
B. Increases
C. Decreases
D. Peaks at approximately 15 MHz

15. Which could you verify by using a spectrum analyzer?
A. Amount of isolation between input and output ports of a VHF duplexer
B. Whether or not a crystal is operating at its fundamental frequency
C. Output spectrum of a transmitter
D. All of the above

16. Which has the greatest affect on frequency counter accuracy?
A. Input attenuator
B. Time base
C. Decade divider
D. Temperature coefficient

17. What can occur in receiver performance if the noise figure is lowered?
A. Reduced signal-to-noise ratio
B. Improved sensitivity
C. Reduced bandwidth
D. Increased bandwidth

18. Which of the following causes intermodulation distortion?
A. Insufficient gain
B. Insufficient neutralization
C. Non-linear circuits and devices
D. Positive feedback

19. What could you do to suppress noise from an electric motor?
A. Install a high-pass filter in series with the motor leads
B. Install an AC-line filter in series with the motor leads
C. Install a bypass capacitor in series with the motor leads
D. Use a ground-fault current interrupter in the motor power circuit

20. Which of the following best describes resonant frequency in a circuit?
A. The maximum frequency that passes current
B. The minimum frequency that passes current
C. The frequency at which capacitive and inductive reactance are equal
D. The frequency at which reactive and resistive impedance are equal

21. To what percentage of the starting voltage will a capacitor in an RC circuit charge after one time constant?
A. 13.5%
B. 36.8%
C. 63.2%
D. 86.5%

22. Which of the following is used for displaying the phase angle of a reactive circuit?
A. Maidenhead grid
B. Faraday grid
C. Elliptical coordinates
D. Polar coordinates

23. What causes resistance in a conductor to differ with direct and RF currents?
A. High frequency pressure on insulation
B. The Heisenburg Effect
C. Skin effect
D. The non-linear nature of conductors

24. Where would you expect to find an excess of free electrons?
A. N-type semiconductor material
B. P-type semiconductor material
C. Superconductor-type material
D. Bipolar-type semiconductor material

25. Which type of diode can be used for both amplification and oscillation?
A. Point contact
B. Zener
C. Tunnel
D. Junction

26. Which of the following is a disadvantage of TTL devices over CMOS logic devices?
A. Limited output capability
B. Higher distortion
C. Prone to damage from static discharge
D. Higher power consumption

27. Which of the following describes a CCD?
A. Utilizes both analog and digital circuitry
B. Suitable for making an audio delay line
C. Samples and stores analog signals
D. All of the above

28. How do you ensure that a crystal oscillator provides the frequency specified by its manufacturer?
A. Supply a specific parallel inductance
B. Supply a specific parallel capacitance
C. Use a specific voltage for biasing
D. Use a specific current for biasing

29. Which part of photovoltaic cells absorb energy from light?
A. The protons
B. The photons
C. The electrons
D. The depletion region

30. Which type of circuit alternates between two states without the use of an external clock?
A. J-K flip-flops
B. T flip-flops
C. Monostable multivibrators
D. Astable multivibrators

31. What type amplifier can get rid of even-order harmonics?
A. Push-push
B. Push-pull
C. Class C
D. Class AB

32. Which are properties of a T-network with series capacitors and a parallel shunt inductor?
A. Low-pass filtering
B. Band-pass filtering
C. High-pass filtering
D. Notch filtering

33. Which of the following are commonly used to set up stable references in linear voltage regulators?
A. Zener diodes
B. Tunnel diodes
C. Silicon-controlled rectifiers
D. Varactor diodes

34. How can you boost the higher audio frequencies in an FM transmitter?
A. Add a de-emphasis network
B. Add a heterodyne suppressor
C. Add an audio prescaler
D. Add a pre-emphasis network

35. What can you use to provide high-stability oscillators in microwave communications?
A. GPS signal references
B. Rubidium-stabilized reference oscillators
C. Temperature-controlled high Q dielectric resonators
D. All of the above

36. How will the gain of an ideal op-amp change with frequency?
A. Increase linearly with increasing frequency
B. Decrease linearly with increasing frequency
C. Increase logarithmically with increasing frequency
D. None of the choices are correct

37. Which can be accomplished with a PLL?
A. AF and RF power amplification
B. Digital input comparison and pulse counting
C. Photovoltaic conversion and optical coupling
D. Frequency synthesis and FM demodulation

38. Which of the following can be communicated with digital waveforms?
A. Audio
B. Video
C. Data
D. All of the above

39. What is deviation ratio?
A. The relation between audio modulating frequency and center carrier frequency
B. The relation between maximum carrier frequency deviation and highest audio modulating frequency
C. The relation between carrier center frequency and audio modulating frequency
D. The relation between highest audio modulating frequency and average audio modulating frequency

40. Which of the following is a spread spectrum technique that alters the center frequency of a conventional carrier in accordance with a pseudo-random list of channels?
A. Frequency hopping
B. Direct sequence
C. Time-domain frequency modulation
D. Frequency compandored spread spectrum

41. What is the typical peak voltage at a standard U.S. household electrical outlet?
A. 120 volts
B. 170 volts
C. 240 volts
D. 340 volts

42. What type of antenna provides no gain?
A. Quarter-wave vertical
B. Yagi
C. Half-wave dipole
D. Isotropic

43. What could happen when you operate a directional antenna at different frequencies?
A. Gain varies
B. E-field patterns reverse
C. H-field patterns reverse
D. Exceed element spacing limits

44. What is the main drawback of placing vertically polarized antennas over imperfect ground?
A. Increased SWR
B. Transposed impedance angle
C. Reduced low-angle radiation
D. Increased low-angle radiation

45. What happens to the gain of an ideal parabolic dish when you double the operating frequency?
A. A 3 dB decrease
B. A 3 dB increase
C. A 6 dB increase
D. No change

46. What is a characteristic of mismatched transmission lines?
A. The reflection coefficient is greater than 1
B. The dielectric constant is greater than 1
C. The SWR is less than 1:1
D. The SWR is greater than 1:1

47. Which has most impact on the velocity factor of transmission lines?
A. Termination impedance
B. Line length
C. Dielectric materials
D. Center conductor resistivity

48. Why would you use a Smith chart?
A. To calculate transmission line impedance
B. To calculate antenna radiation resistance
C. To calculate antenna radiation pattern
D. To calculate radio wave propagation

49. Which standardized power measurement adds system gains and subtracts system losses?
A. Power factor
B. Half-power bandwidth
C. Effective radiated power
D. Apparent power

50. Why are maximum permissible exposure limits separated into electric and magnetic fields?
A. The body reacts to electromagnetic radiation from both
B. Ground reflection and scattering causes field impedance to vary with location
C. Electric and magnetic field intensity peaks can occur at different locations
D. All of the above

ANSWER KEY
1. **D** | 60 meter is the only amateur band with specific transmit channels instead of a frequency range.
2. **C** | Environmental assessments must be submitted to the FCC before placing amateur stations within official wilderness areas and wildlife preserves.
3. **B** | Only auxiliary stations, repeater stations and space stations may automatically retransmit radio signals from other stations.
4. **D** | All amateur class licensees may be the control operator of a space station.
5. **B** | A minimum passing test score of 74% is required to receive an amateur operator license.
6. **C** | The U.S. National Radio Quiet Zone is located in West Virginia near the National Radio Astronomy Observatory.
7. **D** | You can use the Keplerian orbital elements of a specific satellite to calculate its location.
8. **A** | NTSC television transmission occurs at a rate of 30 frames per second.
9. **A** | Receivers suppress radio signals that are not using the correct spread spectrum algorithm.
10. **A** | Baud rate is the number of symbols transmitted per second.
11. **B** | Frequency-shift keying is the most popular data transmission method below 30 MHz.
12. **C** | Earth-Moon-Earth communications are generally less prone to path loss when the moon is at perigee.
13. **D** | Receiving a radio signal from multiple paths often results in an echo.
14. **C** | The max distance of ground-wave propagated signals decrease when frequency increases.
15. **D** | A spectrum analyzer would allow you to verify all of the given choices.
16. **B** | Frequency counter accuracy is most affected by the accuracy of its time base.
17. **B** | Lowering the noise figure can result in improved receiver sensitivity.
18. **C** | Intermodulation distortion is usually caused by non-linear circuits and devices.
19. **B** | Connecting an AC-line filter in series with the motor leads will help suppress noise.
20. **C** | Resonant frequency is achieved when capacitive and inductive reactance are equal.
21. **C** | The capacitor in an RC circuit will charge to 63.2% of the starting voltage after one time constant.
22. **D** | Use polar coordinates to display the phase angle of circuits with resistance, capacitive and inductive reactance.
23. **C** | Skin effect is a tendency for the resistance in a conductor to change when exposed to RF currents.
24. **A** | N-type semiconductor materials contain and excess of free electrons.
25. **C** | Tunnel diodes are special purpose devices used in both amplifier and oscillator circuits.
26. **D** | TTL devices tend to consume more power than CMOS logic devices.
27. **D** | All three choices accurately describe charge-coupled devices.

28. **B** | Using the manufacturer-specified parallel capacitance will ensure a crystal oscillator operates at the correct frequency.
29. **C** | The electrons in a photovoltaic cell absorb energy from light.
30. **D** | Astable multivibrators continually switch from one state to the other and do not require an external clock.
31. **B** | Push-pull amps are capable of reducing and eliminating even-order harmonics.
32. **C** | The capacitors and shunts in a T-network provide high-pass filtering.
33. **A** | Zener diodes are often used to establish stable reference voltages in regulators.
34. **D** | Adding a pre-emphasis network to an FM transmitter will boost higher audio frequencies.
35. **D** | All the choices are suitable for providing stable oscillators in microwave communications.
36. **D** | The gain in an op-amp will not change because of variance in frequency.
37. **D** | Phase-locked loops can be used for frequency synthesis and FM demodulation.
38. **D** | Digital waveforms can be used for audio, video, and data.
39. **B** | Deviation ratio is the relation between maximum carrier frequency deviation and highest audio modulating frequency.
40. **A** | Frequency hopping alters the center frequency of a carrier many times per second.
41. **B** | The typical peak voltage is 170 volts while the typical RMS voltage is 120 volts.
42. **D** | Isotropic antennas have no gain in any direction.
43. **A** | The gain of a directional antenna can vary with operating frequency.
44. **C** | Placing vertically polarized antennas over rocky and otherwise imperfect ground usually results in a reduction of low-angle radiation.
45. **C** | Doubling the operating frequency of a parabolic dish will result in a 6 dB increase in gain.
46. **D** | Mismatched transmission lines typically have an SWR greater than 1:1.
47. **C** | The dielectric materials used in a transmission line will determine its velocity factor.
48. **A** | A Smith chart is used to calculate transmission line impedance.
49. **C** | Station gains and losses are included when determining effective radiated power.
50. **D** | All the choices are correct and applicable to both electric and magnetic fields in regard to maximum permissible exposure limits.

APPENDIX 4
Resource Directory

Web Resources

Federal Communications Commission The FCC regulates interstate and international communications by radio, television, wire, satellite, and cable in all 50 states, the District of Columbia and U.S. Territories. Visit their homepage at: http://www.fcc.gov/

American Radio Relay League The ARRL is the primary representative organization of amateur radio operators to the U.S. Government. Visit their homepage at: http://www.arrl.org/

National Conference of Volunteer Examiner Coordinators The NCVEC is a private organization, and functions to facilitate the intercommunications between the FCC and each VEC. Visit their homepage at: http://www.ncvec.org/

All About Circuits This site provides a series of online textbooks covering electricity and electronics. Visit their homepage at: http://www.allaboutcircuits.com/

ElectronicsTheory.com This site provides a number free lessons on basic electronics theory. It also hosts a selection of computer-oriented lessons, ham radio topics, plus related projects and kits. Visit their homepage at: http://www.electronicstheory.com/

Recommended Reading

ARRL Ham Radio License Manual, 2nd Edition; November 1, 2010; ISBN-10: 0872590976; ISBN-13: 978-0872590977

ARRL General Class License Manual for the Radio Amateur, 7th Edition; April 25, 2011;   ISBN-10: 087259811X; ISBN-13: 978-0872598119

ARRL Extra Class License Manual for the Radio Amateur, 10th Edition; June 25, 2012 ;   ISBN-10: 087259517X; ISBN-13: 978-0872595170

ARRL Handbook for Radio Communications, 91st Edition; October 1, 2013; ISBN-10: 1625950012; ISBN-13: 978-1625950017